Millimeter Wave Antennas for 5G Mobile Terminals and Base Stations

Millimeter Wave Antennas for 5G Mobile Terminals and Base Stations

Shiban Kishen Koul
G. S. Karthikeya

CRC Press is an imprint of the
Taylor & Francis Group, an **informa** business

First edition published 2021
by CRC Press
6000 Broken Sound Parkway NW, Suite 300, Boca Raton, FL 33487-2742

and by CRC Press
2 Park Square, Milton Park, Abingdon, Oxon, OX14 4RN

© 2021 Taylor & Francis Group, LLC

CRC Press is an imprint of Taylor & Francis Group, LLC

Reasonable efforts have been made to publish reliable data and information, but the author and publisher cannot assume responsibility for the validity of all materials or the consequences of their use. The authors and publishers have attempted to trace the copyright holders of all material reproduced in this publication and apologize to copyright holders if permission to publish in this form has not been obtained. If any copyright material has not been acknowledged please write and let us know so we may rectify in any future reprint.

Except as permitted under U.S. Copyright Law, no part of this book may be reprinted, reproduced, transmitted, or utilized in any form by any electronic, mechanical, or other means, now known or hereafter invented, including photocopying, microfilming, and recording, or in any information storage or retrieval system, without written permission from the publishers.

For permission to photocopy or use material electronically from this work, access www.copyright.com or contact the Copyright Clearance Center, Inc. (CCC), 222 Rosewood Drive, Danvers, MA 01923, 978-750-8400. For works that are not available on CCC please contact mpkbookspermissions@tandf.co.uk

Trademark notice: Product or corporate names may be trademarks or registered trademarks, and are used only for identification and explanation without intent to infringe.

Library of Congress Cataloging-in-Publication Data

A catalog record for this book has been requested

ISBN: 978-0-367-44543-0 (hbk)
ISBN: 978-1-003-01026-5 (ebk)

Typeset in Palatino
by SPi Global, India

To

Veena Koul, Mansi Narain, Grace Yoo Koul, Amendra Koul, Anant Koul

and

Shrilalitha Girish, Sadananda G B, Jamuna

Contents

Preface xi
About the Authors xiii
Abbreviations xv

1. **Introduction** 1
   1.1 Need for Millimeter Waves 1
   1.2 Antennas for Cellular Communications 4
   1.3 Contrast between 4G and 5G Architectures 6
   1.4 Antenna Designs for mmWave 5G Mobile Terminals and Base Stations 9
       1.4.1 Antennas for Mobile Terminals 10
       1.4.2 Antennas for Base Stations 12
   1.5 Antennas beyond 5G 13
   1.6 Outline of the Book 13
   References 14

2. **Conformal Antennas for Mobile Terminals** 21
   2.1 Introduction 21
   2.2 Typical Requirements for Mobile Antennas 21
   2.3 CPW-fed Wideband Corner Bent Antenna for 5G Mobile Terminals 23
       2.3.1 CPW-fed Wideband Antenna 24
       2.3.2 CPW-fed Corner Bent Antenna 30
       2.3.3 CPW-fed Corner Bent Antenna with Reflector 34
   2.4 A wideband High Gain Conformal Antenna for mmWave 5G Smartphones 44
   2.5 Design Guidelines for CPW-fed Conformal Antennas at Ka Band 49
   2.6 Conclusion 50
   References 50

3. **Flexible Antennas for Mobile Terminals** 53
   3.1 Introduction 53
   3.2 Overview of Flexible Substrates for mmWave Applications 54
   3.3 Corner Bent Patch Antenna for Portrait Mode 56
   3.4 Corner Bent Tapered Slot Antenna for Landscape Mode 60
   3.5 Dielectric Loaded Polycarbonate-Based Vivaldi Antenna 66
   3.6 Conclusion 71
   References 71

4. **Compact Antennas with Pattern Diversity** 73
   4.1 Introduction 73
   4.2 CPW-fed Conformal Folded Dipole with Pattern Diversity 74
       4.2.1 CPW-Fed Folded Dipole 75
       4.2.2 Conformal Folded Dipole Backed by Reflector 83

	4.3	Conformal Antennas with Pattern Diversity 92
		4.3.1 Mobile Terminal Usage Modes 92
		4.3.2 Conformal Patch Antenna 94
		4.3.3 Conformal Tapered Slot Antenna 96
		4.3.4 Conformal TSA with Parasitic Ellipse 99
		4.3.5 Conformal Pattern Diversity 103
	4.4	Case Studies: Measurement in a Typical Indoor Environment 108
	4.5	Conclusion 110
		References 111

5. Pattern Diversity Antennas for Base Stations 115
   5.1 Introduction 115
   5.2 Pattern Diversity of Path Loss Compensated Antennas for 5G Base Stations 115
       5.2.1 mmWave Tapered Slot Antenna 116
       5.2.2 Dielectric and Metamaterial Loaded TSA 119
       5.2.3 Pattern Diversity 127
   5.3 Path Loss Compensated Pattern Diversity Antennas with 3D Printed Radome 131
       5.3.1 3D Printed Radome for a Patch Antenna 131
       5.3.2 Pattern Diversity with 3D Printed Radome 132
   5.4 Path Loss Compensated Module with Progressive Offset ZIM 136
       5.4.1 Central Element: Tapered Slot Antenna 136
       5.4.2 Spatially Modulated ZIM Loaded Antenna 139
       5.4.3 Stacked Pattern Diversity 141
   5.5 Path Loss Compensated Quasi-Reflector Module 142
   5.6 Design Guidelines for High Aperture Efficiency Antenna 143
   5.7 Case Studies: Measurement in a Typical Indoor Environment 145
   5.8 Conclusion 146
       References 146

6. Shared Aperture Antenna with Pattern Diversity for Base Stations 149
   6.1 Introduction 149
   6.2 Shared Aperture Antenna 150
   6.3 DPZIM Design and Characterization 153
   6.4 Shared Aperture Antenna with DPZIM 154
   6.5 Design Guidelines for High-Gain Dual-Polarized Antenna Module 160
   6.6 Conclusion 160
       References 160

7. Co-Design of 4G LTE and mmWave 5G Antennas for Mobile Terminals 163
   7.1 Introduction 163
   7.2 Miniaturization Techniques for Antenna Size Reduction 163
   7.3 Conformal 4G LTE MIMO Antenna Design 164
       7.3.1 CRLH-Based Conformal 4G LTE Antenna 164
       7.3.2 Compact CRLH-Based Conformal 4G LTE MIMO Antenna 168
   7.4 Conformal mmWave 5G MIMO Antenna 172
   7.5 Corner Bent Integrated Design of 4G LTE and mmWave 5G Antennas 176
       7.5.1 4G LTE Antenna Design 176
       7.5.2 mmWave 5G Antenna Design 177
       7.5.3 Co-Designed Corner Bent 4G LTE and mmWave 5G MIMO Antennas ... 181

Contents

	7.6	Case Study: Co-Design of 4G and 5G Antennas in a Smartphone	184
	7.7	Conclusion	185
	References		186

8. Corner Bent Phased Array for 5G Mobile Terminals 189
    8.1 Introduction 189
    8.2 Phased Array Designs for mmWave Frequencies 190
    8.3 Need for Corner Bent Phased Array 192
    8.4 Corner Bent Phased Array on Polycarbonate 194
    8.5 Design Guidelines for a Phased Array at Ka Band 200
    8.6 Conclusion 200
    References 201

9. Fabrication and Measurement Challenges at mmWaves 203
    9.1 Introduction 203
    9.2 Fabrication Process and Associated Tolerances 203
    9.3 S-parameter Measurements 206
    9.4 Pattern Measurements and Sources of Error 207
    9.5 Gain Measurements 209
    9.6 Conclusion 211
    References 211

10. Research Avenues in Antenna Designs for 5G and beyond 213
    10.1 Introduction 213
    10.2 PCB-Based Antenna Designs for 5G Cellular Devices 213
    10.3 Application of Additive Manufacturing for Antennas 215
        10.3.1 A Dual Band mmWave Antenna on 3D Printed Substrate 216
    10.4 On-Chip Antennas for CMOS Circuitry 219
        10.4.1 A Wideband CPS-Fed Dipole on Silicon 220
    10.5 Optically Transparent Antennas 225
    10.6 Conclusion 226
    References 226

Appendices 229
    Appendix A: Hints for Simulations in Ansys HFSS 229
        A.1 Modelling 229
    Appendix B: Measurement Issues with End-Launch Connector 233
    Appendix C: Material Parameters' Extraction Using S-parameters 234
    Appendix D: Useful MATLAB Codes 235
    References 237

Index 239

Preface

The ever-expanding growth of bandwidth-hungry applications among smartphone users has prompted researchers across the planet to redesign the existing commercial 4G ecosystem to be compliant with future high bandwidth applications. Since, the sub-6 GHz bands are over-crowded, millimeter wave carrier frequencies seem to be one of the promising candidates to ease this congestion. After the famous testing campaign results reported by researchers at New York University, the feasibility of millimeter waves being used as carrier frequencies for cellular communication systems for future 5G is becoming closer to reality.

There are several challenges for the design and deployment of commercial transceiver modules operating at 28 GHz and beyond. Antenna designs are a crucial aspect for the realization of hardware modules at these frequencies, and designs for these antennas are still in the nascent stage. The primary issue is the inherent high path loss at these frequencies, which could be mitigated by high gain antennas on both the mobile terminals and base stations. But the solution to realize this objective might not be as obvious as it seems. The special design requirements of antenna designs with millimeter waves for mobile terminals and base stations offers unique challenges to be addressed.

The context of millimeter wave 5G is introduced initially in this book. The design requirements for mobile terminals is laid out, and a couple of design examples are also illustrated with sufficient technical depth and justification. The typical data usage modes of future 5G mobile terminals are introduced, and the challenges associated with the design of antenna modules to cater to these needs is also described. A real-world deployment scenario is also investigated to prove that the proposed antenna modules work in an integrated prototype smartphone despite the multipath effects in an indoor environment.

Design requirements for future millimeter wave base stations is discussed, followed by the introduction of the concept of path loss compensation. Compact pattern diversity modules are investigated to meet these design requirements. Design principles of high aperture efficiency antennas in the context of tapered slot antennas are outlined, and a dual-polarized compact pattern diversity module is described for base station application for high throughput.

Next, the need for the co-design of commercial 4G antennas with millimeter wave 5G antennas is explained. The standard miniaturization techniques to reduce the antenna size is discussed. A conformal 4G antenna with a low physical footprint is proposed. MIMO topology of the 4G antenna is also designed and characterized. A conformal millimeter wave antenna operating in the 28 GHz band is also reported. In a case study, the integration of the low frequency 4G antenna with the mmWave 5G antenna is explained in detail. The need for phased array modules in the context of 5G mobile terminals is introduced next in the book. A comparison of phased array versus pattern diversity modules is presented. Phased array designs in the context of millimeter waves is explained in detail, including a section on beam scanning for a multi-port antenna system. The need for a corner bent array is explained, along with the characterization of planar and corner bent transmission lines on a flexible substrate. Design constraints, along with the characterization of a corner bent phased array designed on polycarbonate substrate, are also included in the book. Basic guidelines to design a phased array at millimeter wave frequencies are then given. The book also includes one chapter each on Fabrication and measurement challenges at mmWaves, and Research avenues in antenna designs for 5G and beyond.

Authors

Shiban Kishen Koul earned a BE degree in electrical engineering from the Regional Engineering College, Srinagar, Jammu and Kashmir, India, in 1977; an MTech in radar and communication engineering in 1979; and a PhD in Microwave Engineering in 1983 at the Indian Institute of Technology (IIT), Delhi. He served as Deputy Director (Strategy & Planning) from 2012 to 2016, and Dr R. P. Shenoy Astra Microwave Chair Professor from 2014 to 2018 at the IIT Delhi. He is presently the Deputy Director (Strategy & Planning and International Affairs) at the Indian Institute of Technology, Jammu. He also served as the Chairman of M/S Astra Microwave Products Limited, Hyderabad, a major company involved in the development of RF and microwave systems in India, from 2009 to 2019. His research interests include RF MEMS, high frequency wireless communication, microwave engineering, microwave passive and active circuits, device modelling, millimeter wave IC design, and reconfigurable microwave circuits including antennas. He has successfully completed 36 major sponsored projects, 52 consultancy projects and 58 technology development projects. He is author/co-author of 450 research papers, 10 state-of-the art books, and 3 book chapters. He holds 11 patents and 6 copyrights. He is a Life Fellow of the Institution of Electrical and Electronics Engineers, USA (IEEE), Fellow of the Indian National Academy of Engineering (INAE), and Fellow of the Institution of Electronics and Telecommunication Engineers (IETE). He is the Chief Editor of the *IETE Journal of Research* and Associate Editor of the *International Journal of Microwave and Wireless Technologies*, Cambridge University Press. He has delivered more than 280 invited technical talks at various international symposia and workshops. He served as the MTT-S ADCOM member from 2009 to 2018, is a Member of IEEE MTT Society's technical committees on Microwave and Millimeter Wave Integrated Circuits (MTT-6) and RF MEMS (MTT-21), and of the India Initiative team of IEEE MTT-S, Adviser Education Committee, Membership Services Regional Co-coordinator Region-10, member of the Sight Adhoc Committee MTT-S, and an MTT-S Speaker Bureau lecturer. He was a distinguished microwave lecturer at IEEE MTT-S for the period 2012–2014, and a distinguished microwave lecturer emeritus at IEEE MTT-S in 2015.

He is a recipient of a Gold Medal from the Institute of Electrical and Electronics Engineers, Calcutta (1977); the S. K. Mitra Research Award from the IETE for the best research paper (1986); Indian National Science Academy (INSA) Young Scientist Award (1986); International Union of Radio Science (URSI) Young Scientist Award (1987); the top Invention Award (1991) of the National Research Development Council for his contributions to the indigenous development of ferrite phase shifter technology; VASVIK Award (1994) for the development of Ka-band components and phase shifters; the Ram Lal Wadhwa Gold Medal from the Institution of Electronics and Communication Engineers (IETE) (1995); an Academic Excellence award (1998) from the Indian Government for his pioneering contributions to phase control modules for Rajendra Radar; the Shri Om Prakash Bhasin Award (2009) in the field of Electronics and Information Technology;

a Teaching Excellence award (2012) from IIT Delhi, an award for his contribution to the growth of smart material technology (2012) by the ISSS, Bangalore; the Vasvik Award (2012) for contributions made to the area of Information and Communications Technology (ICT); the M. N. Saha Memorial Award (2013) from the IETE for the best application-oriented research paper; and the IEEE MTT Society Distinguished Educator Award (2014).

G. S. Karthikeya gained his undergraduate degree in electronics and communication engineering in 2010 from the Visvesvaraya Technological University, Belgaum. He received a Master's degree in microwave engineering from the University of Kerala in 2012. He worked as an Assistant Professor in Visvesvaraya Technological University from 2013 to 2016, where he established the Antenna Architects' lab. He joined the Centre for Applied Research in Electronics, IIT Delhi in January 2017 and defended his thesis in December 2019. He has authored or co-authored more than 40 articles in peer-reviewed journals and conference proceedings. He has also filed three Indian patents and two US patents. His research interests include metamaterials, EBG structures, and mmWave antennas for mobile terminals and base stations. He is a member of the IEEE Antennas and Propagation Society, and the Antenna Test & Measurement society. He serves as the reviewer of ACES, IEEE Access and Cambridge's IJMWT. He has participated actively in more than 15 workshops on antennas in India and abroad.

Abbreviations

1G	first generation
2G	second generation
3G	third generation
4G	fourth generation
5G	fifth generation
ADS	Advanced Design System
AUT	antenna under test
AVA	antipodal Vivaldi antenna
CAD	computer aided design
CDMA	Code Division Multiple Access
CMOS	complementary metal- oxide semiconductor
CPS	coplanar stripline
CPW	coplanar waveguide
CRLH	composite right/left-handed
CSRR	circular slot ring resonator
CST	computer simulation technology
CVD	chemical vapour deposition
DPZIM	dual-polarized zero-index metamaterial
EBG	electromagnetic band-gap
ECC	Envelope Correlation Coefficient
ERV	effective radiating volume
FDM	fused deposition modelling
GDP	Gross gross domestic product
GPS	Global Positioning System
HFSS	High Frequency Structure Simulator
HSPA	High Speed Packet Access
IBW	impedance bandwidth
LCP	liquid crystal polymer
LOS	line of sight
LPDA	log-periodic dipole antenna
LTE	Long Term Evolution
MIMO	multiple input, multiple output
mmWave	millimeter wave
MTM	metamaterial
NLOS	non-line of sight
PCB	printed circuit board
PEC	perfect electric conductor
PET	polyethylene terephthalate
PLA	polylactic acid
PN	pseudo-noise
PR	photo Rresist
QWT	quarter-wave transformer
RF	radio frequency

RFIC	radio frequency integrated circuit
SAR	Specific Absorption Rate
SIW	substrate integrated waveguide
SMA	Sub-Miniature version A
SMS	Short Messaging Message Services
SMZIM	spatially modulated zero-index metamaterial
TEM	transverse electromagnetic mode
TSA	tapered slot antenna
UAV	unmanned aerial vehicle
UV	ultraviolet
VNA	vector network analyzeranalyser
WiFi	wireless fidelity
ZIM	zero-index metamaterial
ZOR	zeroth order resonator

1
Introduction

1.1 Need for Millimeter Waves

Marconi demonstrated the first transatlantic wireless transmission in the early 1900s, marking the beginning of wireless technology. The data transmissions in the early days were concerned with binary messages and gradually evolved to become advanced data transmission. This evolution in turn meant that the carrier frequency and its associated bandwidth had to be increased. The explosion of wireless devices is primarily a result of the wireless revolution that started in the 1980s. The first generation cellular was primarily analogue in nature, needing crude resonant antennas for communication. It involved simple frequency modulated schemes for data transaction, so modulators and demodulators had to be designed to accommodate this new frequency conversion scenario. The purpose of frequency upconversion was to achieve a feasible antenna size on both the base station and the mobile handset. It is interesting to note that the antennas on the 1G mobile handsets were also electrically small with poor radiation efficiency, a trend observed even in modern-day smartphone antenna design. After a decade of 1G, 2G was deployed with digital encoding in 1992. The switch from analogue based communication to a digital ecosystem was one of the pioneering events witnessed in the industry. The bandwidths and data rates were severely limited by the analogue communication network, primarily because of a lack of sophisticated analogue signal processing schemes to deal with analogue data. Even though the back-end electronics supported digital signal processing in the 2G era, the front-end transmission would always be an analogue signal, since the antennas would not be able to transmit a digital signal per se. Hence the drama of digital to analogue conversion and digital modulation schemes are designed such that the signals are antenna-transmission friendly. The antenna requirements were not very critical even at this stage since the mobile phones of the era were still not easily portable; for example, the 2G phones were almost 30% larger compared to the commercial smartphones of the present day. The antennas were usually telescopic, meaning they had to be expanded when the user wanted to make a call, as the length of the monopole or helical antenna was twice that of the mobile handset for the prescribed frequency of operation. The choice of frequency of operation is a global phenomenon, where experts from various nations would agree on a certain set of frequencies. Based on the agreed frequencies, hardware designers created radios centred around them, with strict bandwidth requirements. The manufacturers would generally design cellular equipment to accommodate as many standards as possible to keep the radio both backward compatible and future proof. 3G was introduced in 2001, offering higher data rates to users, with video streaming and other fancy services on compact mobile phones. The arrival of video data on the smartphone was received extremely well by most users. Video is considered to be the epitome of digital data compared to plain and uninteresting voice data. Video data transmissions meant that the spectral efficiency had to be very high to achieve a decent transmission to the user without annoying delays and buffering. High speed transceivers had to be built for video transmission and reception. At this stage, antennas on the mobile phone had to be compact, with stringent design requirements, because several bands with specific bandwidths were allocated in the 3G standards. Even though 3G promised high data rates and video transmission on the menu, it was not very well received by the market, which led to the design and evolution of 4G LTE (Long Term Evolution) in 2011 with even higher data rates to subscribers. Antennas designed for these systems had to be designed with backward compatibility, thus making it even more challenging to achieve. Meanwhile, smartphones were becoming multi-featured and multi-functional, with several wireless chipsets in addition to the typical cellular chipsets to be accommodated in a severely constrained space inside a mobile terminal. WiFi, Bluetooth and GPS chipsets were to be designed, along with conventional cellular standards. Table 1.1 gives an overview of the cellular evolution [1]. As observed in the table, 1G had no official requirements and the antenna requirements were simple. A monopole with retractable design would be sufficient, since bandwidth and the gain were not major concerns. Wireless evolution suggests that, as the generation progresses, carrier frequency and its corresponding bandwidth allocation also increases. The transition to digital telephony happened in the 1990s, when the concept of short messaging services (SMS) was introduced. The SMS service could be brought to subscribers through the storage and processing of digital data. Internet services could be accessed by users but with very limited data rates, thus forcing them to use low data rate applications such as text. As the wireless standards progressed to 3G, the international committee laid out necessary data rates to be supported by both manufacturers and service providers. 4G experienced phenomenal data growth and a huge market success primarily as a result of the introduction of video streaming services.

Historically, wireless standards typically evolve in a decade, but the same trend may or may not continue in the future. The scepticism regarding 5G is because both manufacturers and service providers have to redesign both the hardware and the network architecture. Previous generations had similar carrier frequencies, which meant that the migration towards the next generation was minimal compared to the proposed migration from 4G to 5G. According to Cisco's projections, by 2021 global mobile data traffic will reach 49 exabytes (EBs) per month, 4G connections will be 53% of total mobile connections and will account for 79% of mobile data traffic [2]. It is predicted that the mobile telephone industry would contribute 4.2% of global GDP by the end of the 2020s, which is a significant share compared to other sectors. Most of the subscribers on the mobile network utilize it for

TABLE 1.1

Evolution of Cellular Technologies

Generation	Requirements	Remarks
1G	No official requirements, mainly analogue system	Deployed in 1980s
2G	No official requirements, transition to digital technology	Deployed in 1990s. SMS and low data rate
3G	IMT-2000 required 144kbps mobile, 384 kbps for pedestrians, 2 Mbps for indoor	Deployed in 2001 CDMA and HSPA
4G	40 MHz channels with high spectral efficiency	Deployed in 2011 LTE-Advanced

Introduction

bandwidth-hungry applications such as video streaming, online gaming and video calling. It is interesting to note that the users would be encouraged to utilize these services more often using the existing system. To cater to the needs of this massive explosion of data consumption, innovative radio designs would be essential in the hardware ecosystem, communication protocol design and cellular network layout. Pundits in the industry predict that the evolution from 4G cellular telephony to 5G would be an arduous task, since 5G needs a revolutionary restructuring of the network to cater to the needs of future consumers. It also depends on the preferences of the designers and committee heads in allocating frequency of operation and associated bandwidth.

Research and development (R&D) investment for 5G design, development and deployment has increased in recent years: for example, the European Commission (EC) has invested €50 million to date; the METIS project is an outcome of that funding. Numerous academic institutions such as the University of Surrey, TU Dresden, Lund University, etc. have established innovation centres targeted at 5G architecture design. The Chinese government has initiated IMT-2020 to promote and standardize research in 5G wireless technologies. The Koreans have promoted a 5G forum. Leading companies such as Samsung, Qualcomm, Ericsson, Verizon, Nokia Siemens networks, NXP, NTT Docomo, IBM, and Huawei have established dedicated 5G research groups. The Government of India has also set up 5G research groups across the country to help with the 5G standardization process.

Since, the sub-6 GHz bands have high spectral congestion, future wireless standards could be migrated to the millimeter wave frequency range, which is easier said than done, as the design for sub-6 GHz bands is well-known for operating link budgets and receiver sensitivity. But the inherent problem with the millimeter wave frequencies is the high path loss compared to existing cellular communication standards. It must also be observed that the atmospheric absorption in the 60 GHz band is close to 20 dB/km, as is evident from Figure 1.1. An attenuation of 7 dB/km is observed as a result of heavy rainfall rates of 1 inch (25 mm) per hour for cellular propagation at 28 GHz, which translates to a mere 1.4 dB of attenuation over 200 m distance. Free space path loss and the penetration losses are

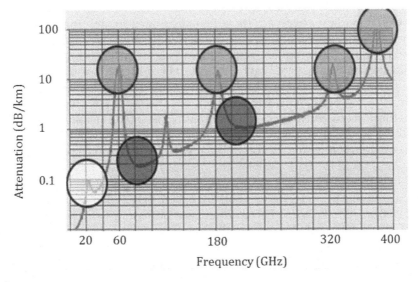

FIGURE 1.1
Atmospheric absorption across the frequency spectrum [7].

relatively higher, which could be mitigated by high gain antennas. Penetration loss could be as high as 30–40 dB for common building materials such as bricks, concrete and glass window panes. This aspect would deteriorate the link budget severely, the feasibility of maintaining the link budget with decent power at the base station or access point for a given receiver sensitivity, is yet to be implemented.

1.2 Antennas for Cellular Communications

The antennas designed in the previous wireless generations for mobile terminals could be broadly classified as external and internal antennas [3]. The external antennas appeared in the older wireless generations because of the lower operating frequency and poor receiver sensitivity. Eventually, antennas were integrated with the mobile device, where the geometry of the designed antenna would fit snugly inside the typical mobile case, sacrificing the gain and radiation efficiency of the internal antenna but the receivers would still be operational as a result of advances in receiver design and digital communication protocol changes. The advanced digital communication strategies proposed led antenna designers to sacrifice the gain and effective radiated power from the antenna. The designer just had to make sure that the antenna could be manufactured with the industry standard techniques and operate at the specified band. Even the impedance matching between the antenna and the receiver chain was also not very critical.

During the first generation of mobile telephony, the lower carrier frequency forced the designers to incorporate resonant antennas which were inherently narrowband, consequently forcing the subscribers to use only voice-related services. These antennas were typically telescopic whip antennas which could be extended during the call [4]. The balun that acted as a matching network was integrated with the circuitry within the phone, but the metallic whip antenna was exterior to the phone. The technique of achieving impedance matching of the electrically small antenna with the receiver chain using a matching network is still an industry standard practice.

As the wireless standards evolved, the antenna designs for the mobile terminals also evolved in conjunction because of the accommodation of multi-functional chipsets operating in different bands. As the frequency of operation was increased to 900 MHz during the 2G wireless standards, it meant that antennas could be designed with a physical size that could fit inside a typical mobile case. This is illustrated in Figure 1.2. The antenna was incorporated into one of the panels enclosing the device. The mobile device had liquid crystal display on the front along with the keypad. Beneath these modules was the motherboard which housed the RF front end and the digital processing unit. The motherboard would also be connected to the massive lithium ion battery. Hence the only logical placement of the antenna would be on the back panel of the casing. The connector pins would also have an additional matching circuit before feeding into the actual radiating structure.

The antennas would be electrically small such that they could be integrated with the motherboard of the mobile phone. These antennas would offer low radiation efficiency in the range of 5%–10%, primarily because of their electrically small size. The substrate used for the motherboard design has high dielectric constant and forces the antenna also to use an identical substrate, thus leading to additional deterioration in radiation efficiency. An antenna once placed with the motherboard loses its effective radiated power as a result of attenuation by co-located modules such as cameras, speakers, microphones etc. The net

Introduction

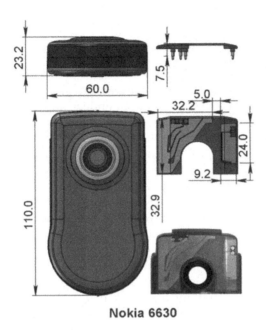

FIGURE 1.2
Antenna of a popular commercial mobile phone with GSM and 3G [3].

radiated power of the mobile handset is further decreased by the user when the mobile terminal is operational for data usage or on a call. Since the path loss at the sub-6 GHz frequencies is relatively low, a communication link is definitely feasible with these antenna specifications. The bandwidth of these antennas must be compliant with the promised wireless standards within which the mobile phone is supposed to operate. The compliance and certification agencies are very strict regarding bandwidth constraints. The out-of-band emissions by the antenna must be within specified limits as detailed by these standardized agencies, and sometimes additional design tweaks are essential to eliminate harmonic and spurious emissions. Thus, the antenna must not only be multi-band but also have adequate bandwidths in the corresponding bands. The desirable pattern of the mobile integrated antenna would be hemispherical, wherein the radiation is mainly away from the head. The gain of the integrated antennas would typically be close to 1–2 dBi, given the electrical size of the antennas. It is important to design antennas which offer the lowest Specific Absorption Rate (SAR), hence the placement of the antenna in the mobile case is an equally important design constraint.

The antenna could never be designed as a stand-alone entity; it has to be co-designed with the transceiver chain. Hence the input impedance of the antenna must be matched to the power amplifier, low noise amplifier and the switch. The leakage from transmitter to receiver must not be contributed by the antenna's poor impedance match. Antenna post-placement with other entities on the mobile terminal must also be studied to understand the deterioration in the antenna's characteristics, since metallic or quasi-metallic modules near the motherboard of the phone tend to alter the antenna's operational characteristics. Also, the effects of the user's hands must be taken into consideration when designing the antennas, since the user's hand tends to detune the antenna. The trend in current commercial smartphones is to place the antenna within a metal rim, which makes it even more challenging to design the antenna.

1.3 Contrast between 4G and 5G Architectures

The typical layout for a 4G cellular network is demonstrated in Figure 1.3. To illustrate the cellular layout, 900 MHz is chosen as an example. It is clear that the base stations are located at 1km from each other and the cellular boundary is at close to 500 m. The density of the base station antennas would be decided by the actual number of users, but the generic layout presented here serves as a reference [5]. The scenario presented here would work in the context of a free space, obstruction-free environment. In an urban setting, the base station placements would be optimized to accommodate the estimated number of subscribers in the corresponding cell. Layout designers would also add a margin to accommodate the growth of subscribers in a given cell. The layout presented here would be a single operator, but in reality, multiple operators would provide services in a given urban setting, making the situation even more complicated for actual layout design. These base stations would also inherently have point to point links to other base stations for high data-rate exchange among them. These point to point links use high gain high frequency antennas to achieve this. The base station antennas so designed would cover the ground omni-directionally. The placement of the base station antennas also depends on the spatial arrangement of the mobile terminals. For example, if the urban setting has too many multi-storeyed buildings or skyscrapers, then the base station antennas must be mounted in an elevated position to give adequate coverage. In contrast to the 900 MHz layout, the 28 GHz carrier signal's cellular layout is also shown in Figure 1.3. Since the path loss is higher than that of its 4G counterpart, the distance between the base stations has been decreased from 1 km to 400 m to maintain an identical link budget. The proposed layout is applicable when the 28 GHz carrier signal is being used in an outdoor environment, as the penetration losses are higher than the traditional wireless links. The ideal scenario for cellular layout for mmWave 5G is 200 m distance between the base stations, which is an over-simplification, as buildings and various other obstacles would also lead to heavier

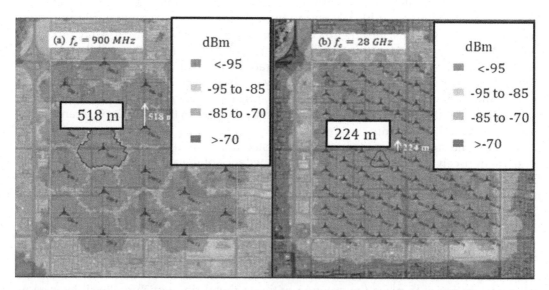

FIGURE 1.3
Cellular network for 4G and proposed 5G [5].

Introduction

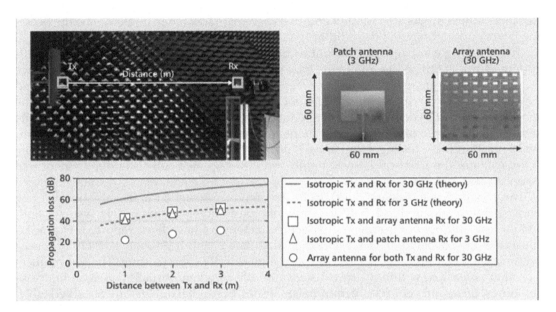

FIGURE 1.4
Comparison of Propagation loss and gains for 4G and 5G [6].

attenuation in addition to the path loss. Multi-path interference effects are also not very well understood in this frequency domain.

Since the path loss for mmWave 5G is higher than that of 4G systems, in order to maintain the link budget it is essential to incorporate higher gain antennas on both the base stations and mobile terminals, as demonstrated in [6]. The physical footprint of the antennas remains identical for 3 GHz and 30 GHz to achieve a decent link budget, as illustrated in Figure 1.4. Even though researchers claim that high gain antennas could easily be incorporated on both radios of the cellular communication link, the feasibility of this is yet to be investigated thoroughly. High gain antennas on the base station would be easier to implement, since the physical aperture is not a limiting criterion. But high gain antennas on the mobile terminal would mean that the electrical aperture of the antenna has to be large enough to achieve the desired gain. The consequence of increased gain is that the beamwidth would be decreased, hence deteriorating the coverage. Experts opine that beam locking could be done by implementing phased arrays on both the mobile terminal and the base station. This would work well for the prescribed link budget for a couple of users, but as the simultaneous number of users increases in the given vicinity the complexity of the beam locking and control signals, along with the processing for angle of arrival, would also increase. This would mean that the regular base stations have to be replaced by smart base station antennas that would lock on to the signal for the user as well as providing support for simultaneous users in the context of multi-beam.

Researchers at New York University (NYU) have performed exhaustive channel propagation studies at 28 and 38 GHz bands to prove the feasibility of mmWave bands for cellular telephony [7]. According to this study, there are many parameters that hinder propagation at 28 GHz, such as path loss, rainfall, penetration losses etc. An attenuation of 7 dB/km is observed as a result of heavy rainfall rates of 1 inch (25.4 mm) per hour for cellular propagation at 28 GHz, which translates into a mere 1.4 dB of attenuation over 200 m distance. Which proves that rainfall and free space path loss alone might not be the deciding factors when designing the cellular link.

In the testing campaign, a transmitter horn antenna with a gain of 24.5 dBi and 10° beamwidth was utilized to mimic the base station antenna. This might be feasible, since an electrically large aperture could be designed at the base station with a transmitted power of at least 30 dBm. A horn antenna with similar specifications was used for the receiver. A gain of 24 dBi would be challenging to achieve with the aperture at the mobile terminal, and hence phased array systems are proposed in various articles. The beamwidth used in these testing campaigns is of the order of 15°, translating to very narrow coverage, and the trade-off between gain and coverage in a full-scale deployment scenario is yet to be seen.

The penetration losses are 20–40 dB for common construction materials such as brick and concrete, compared to 1–5 dB for conventional commercial sub-6 GHz systems. Both indoor and outdoor environment channel characteristics have been reported. Figure 1.5 illustrates the coverage of 28 GHz carrier modulated with high data rate PN sequence within a 200 m radius in New York, a typical high-density urban setting. The communication link is feasible within a short distance of 200 m when obstacles such as high-rise buildings, trees, and vehicles are minimal or absent, but signal outages are observed when the density of the obstacles increase in the given cellular area. It must also be observed that the line of sight (LOS) link has a higher received power compared to the non-line of sight (NLOS) scenario, since the NLOS situation would be the composite signal arriving from multiple reflections from different obstacles, and the NLOS signal cannot be modelled accurately because of the intricate nature of the surroundings. Hence the primary

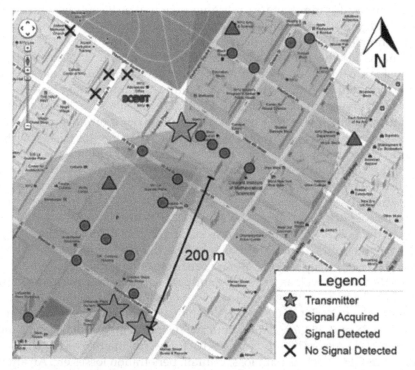

FIGURE 1.5
Coverage of 28 GHz signal in New York [7].

Introduction

TABLE 1.2

Comparison of 4G and 5G Antennas for Cellular Communication

Parameter ↓	4G	Proposed 5G
Frequency	0.7–GHz	28 GHz
Free space Path loss at 200 m	87 dB	107 dB
Penetration losses	5–10 dB	20–40 dB
Antenna in mobile terminal	Electrically small Low gain	Phased array? Pattern diversity?
Antenna in base station	Phased array with fixed beam tilt	

challenge to design and deploy a 28 GHz cellular bi-directional communication system is to mitigate free space path loss and penetration losses at the 28 GHz band with an optimal link budget with feasible antenna systems at base station and the mobile terminal. The indoor channel response specifically at 28 GHz is reported in [8], which illustrates a deployable path loss model. Several testing cases for indoor environments are also presented in this book to confirm the validity of millimeter waves as a potential candidate for indoor links.

Antennas for both the mobile terminal and base station are an important aspect of 5G hardware research, in addition to the chipset design and other analogue signal processing hardware. Table 1.2 illustrates the important aspects of the current commercial 4G antennas and the proposed 5G antennas. The current 4G operates in the 0.7–2.7 GHz band with a free space path loss of 87 dB at 200 m in contrast to the proposed 5G, centred at 28 GHz, which faces an additional 20 dB of path loss, and thus needs compensation, as mentioned in section 1.2. The penetration losses are higher in the mmWave 5G regime compared to 4G. This is a much more interesting problem to tackle compared to the path loss. 4G antennas on mobile terminals are electrically small and low gain with high angular coverage, but the same design philosophy would fail if translated to 5G mobile terminals as low gain would be creating an infeasible link budget. 4G base station antennas have high gain with fixed beam tilt illuminating the ground, also the base station antennas have dual polarized antennas to also achieve polarization diversity. The base station antennas for the proposed 5G also need to be high gain, but in order to achieve reasonable angular coverage phased arrays or pattern diversity could be designed. Phased arrays offer beam scanning without blind spots for a limited scan angle, typically ±60°, but would fail when the beam is supposed to be excited from orthogonal directions, hence pattern diversity might be a solution here.

1.4 Antenna Designs for mmWave 5G Mobile Terminals and Base Stations

The typical layout of a generic smartphone is illustrated in Figure 1.6, as explained in [9]. The motherboard occupies most of the space. The void in the motherboard is primarily to mount the battery. The antennas to be integrated with the phone must be compliant with these dimensions. As a rule of thumb, the mobile terminal antenna must fit the dimensions

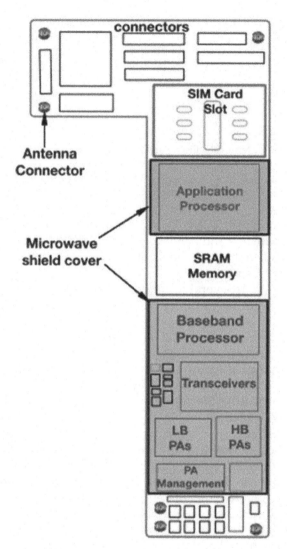

FIGURE 1.6
Internal layout of a typical smartphone [9].

of 10 cm × 5 cm × 0.5 cm. In addition, the 4G antennas and other antennas corresponding to the older wireless standards need to be accommodated within these panel dimensions. The motherboard, along with the battery, is enclosed in an RF shield, a sheet of aluminium. Industry experts indicate that antenna designers must not meddle with the RF shielding enclosing the motherboard as that would compromise the signal integrity of the motherboard.

1.4.1 Antennas for Mobile Terminals

Several articles have been published specifically targeting the mobile terminal application operating at 28 GHz, the rise in published research on millimeter wave antennas coming after the successful testing campaign reported by Prof. Ted Rappaport and the pioneering phased array designs published by Samsung. One of the most popular approaches for mmWave 5G antenna on mobile terminals is phased arrays, as illustrated in [10–15]. The phased array of the antenna element is mounted strategically on the mobile terminal to provide an optimum link irrespective of the orientation with respect to the user. The problem of decent received power irrespective of the user's orientation is unique to the millimeter wave as the cellular antennas on older wireless radios used an omnidirectional antenna, and the user's orientation with respect to the base station's beam would be irrelevant. The common problem with a phased array is the beam widening effect when the beam is more than 40° away from the bore sight, commonly called scanning loss. The effects of the user's hands are characterized in [16–18] and it has been demonstrated that hands would indeed create additional attenuation. Thus phased arrays integrated on a mobile terminal would face scanning loss and finger blockage, which has to be compensated for in the design. Phased arrays with parasitic elements for compaction are reported in [19,20]. Usually, parasitic elements would be frequency sensitive and would lead to narrower bandwidths.

A wideband cavity slot antenna fed with a microstrip to SIW transition is demonstrated in [21]. The advantage of this design is the reduced insertion loss as a result of the incorporation of the quasi-waveguide mode. The fabrication cost is relatively higher due to the series of vias to be integrated into the dielectric. SIW based designs reported in [22–24] have high radiation efficiency but would increase the manufacturing cost. Also, the production line must be tweaked to incorporate the plated through hole technique and would depend on existing production methods. If similar characteristics could be achieved with conventional PCB manufacturing techniques, then the design would be feasible for production-line use.

To achieve diversity for multi-orientation of the smartphone, pattern diversity is another alternative to phased arrays, which reduce the number of controllers for the operation. Also, designing wideband phase shifters is a challenging problem, and co-design of radiators and phase-shifters must be addressed. One of the topologies to achieve pattern diversity is to design reconfigurable antennas, as suggested in [25–28], but these designs operate in the sub-6 GHz and designing the same at Ka band is challenging. A standard quasi Yagi antenna with pattern reconfigurability is illustrated in [29], the four high gain end-fire antennas with parasitic directors would be able to switch the beams in four orthogonal directions but the diode realization is not presented. A more detailed analysis of the diode design is described in Chapter 10.

Antennas designed on flexible substrates are presented in [30–38]. The micromachined designs have lossy substrates reducing the gain, and the PET film-based substrate also has relatively higher loss tangent deteriorating the gain of the antenna; in addition, PET films are expensive compared to polycarbonate. For an example, a CPW fed over-moded slot antenna on low cost flexible PET substrate with a bandwidth of 22–40 GHz and broadside gain of 5–8 dBi is presented in [39].

The antennas designed to target mobile terminal applications would eventually be integrated within a metallic casing, as observed in most of the commercially available smartphones. The antenna designs reported in [40–43] illustrate these effects. The parasitic effects exclusively from the metal rim must be investigated thoroughly to understand the detuning effects of the metal rim on the antenna.

The mmWave 5G antennas must be integrated with existing 4G antennas, since backward compatibility would be an essential feature of future smartphones. Co-designs reported in [44,45] illustrate the challenges and remedies to achieve this. The mutual coupling between 4G antennas with pattern diversity operating at 2.2 GHz and the 28

GHz antenna is characterized. It must be noted that there are few reported articles on the co-existence of 4G and 5G antennas with a large ground plane. The design philosophy of the co-design of a 4G LTE antenna with millimeter wave 5G is explained in detail in Chapter 7.

The designs in [46,47] are complete package of the phased array antennas integrated with narrow band phase shifters. The CMOS chip reported by the IBM research group [48] is a complete transceiver chip which operates at 28 GHz with beam scanning (± 50°) and gain control. The phase shifter along the control architecture was integrated into the chip. The designs described are compatible with the industry standard process flow.

1.4.2 Antennas for Base Stations

Antennas specific to base stations are not commonly reported in the literature. Coplanar Stripline CPS fed antennas are demonstrated in [49–51]. The disadvantage with these designs is that the feed must be re-designed to match the standard microstrip feed. But the radiation patterns are stable across a wide band. Log-periodic dipole antenna designs of [52–54] require empirical relationships for their design in spite of the higher gain for minimal footprint. A tilted parasitic is proposed in [55] for gain enhancement but the pattern integrity is low across the band.

Various research articles on dense dielectric patch arrays have been proposed in [56,57]. The antenna array reported in [56] incorporates a multi-layered design feeding high dielectric constant (ε_r = 82) resonator antenna. Another superstrate is also integrated to the primary radiator to enhance the gain in broadside. Since the aperture coupling was designed with a standard dielectric substrate, EBG unit cells were integrated periodically to suppress the surface wave modes, thus increasing the radiation efficiency.

Leaky wave antennas have been demonstrated in [58,59], where the beam tilt is observed for a shift in the frequency of operation, which might be unsuitable for commercial deployment. Also, leaky wave antennas suffer from scanning loss when the beam is scanned away from the boresight.

Side-lobe suppression techniques have been proposed in [60–62] by parasitic patch and cleverly designing the aperture for phase-error correction when the beam is tilted away from boresight.

Base station designs with PIN diodes incorporated in the aperture is demonstrated in [63–67]; the PIN diodes would be operational in the sub-6 GHz band but proving to be unsuitable at 28 GHz. Circularly polarized designs are proposed in [68–76] but they suffer from narrowband.

A printed ridge gap waveguide feeding method is proposed in [77,78]. The surface waves are minimal in this architecture because of a lack of dielectric between the top metal plane and the ground plane, which functions as an electromagnetic bandgap structure.

A series-fed array with low sidelobe level is proposed in [79,80]. Even though the patterns and broadside are stable in the band 24.5–30 GHz, the waveguide to microstrip transition increases the complexity of the design.

All metallic tapered slot antenna with an impedance bandwidth of 22.5–32 GHz is proposed in [81], which is illustrated in Figure 1.7. The insertion loss is minimal because of the absence of lossy dielectrics. The end-fire gain of the compact 1 x 4 array is 12–15 dBi. The transition from the end launch connector to the all-metallic structure would be challenging to design. Variants of tapered slot antennas are proposed in [82–84]. A similar tilted tapered slot antenna is reported in [85]. The patterns are stable within the band. Co-existence with 4G base station antenna operating at 2–6 GHz is also investigated.

Introduction

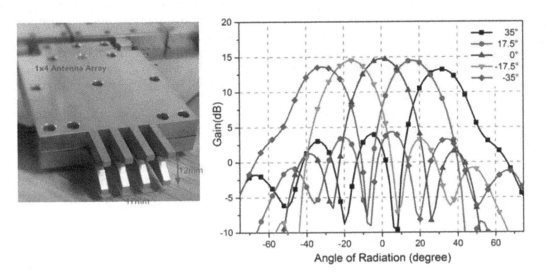

FIGURE 1.7
Phased array design along with beam scanning plot [81].

1.5 Antennas beyond 5G

As the wireless standards evolve it is believed that the carrier frequency would also increase accordingly. For devices beyond 5G, researchers are designing sub-systems beyond 100 GHz, but conventional PCB based circuitry would incur high insertion losses at these frequencies, hence technologies such as Polystrata® is used for micromachining in silicon, as demonstrated in [86]. Here, thin layers of copper traces would be deposited sequentially over layers of dielectric substrate. The entire fabrication process is compliant with the conventional CMOS fabrication, hence leading to higher throughput when mass-produced.

1.6 Outline of the Book

Chapter 1 introduces readers to the scope and relevance of 5G cellular networks, followed by an exhaustive literature survey of reported articles specific to mmWave mobile terminals and base stations. The features and the research gaps of the corresponding designs are also elaborated.

Chapter 2 presents two conformal antenna designs targeting mobile terminals. The first is a conformal wideband, high gain element with a CPW feed. The second proposed element is also a wideband element, with high pattern integrity across the band. Design guidelines to design corner bent antennas in the mmWave band are also discussed.

Chapter 3 introduces flexible antenna designs for compact portable devices. Two conformal antennas are introduced with design details operating at 28 GHz, and an overlapped architecture offering orthogonal pattern diversity is also explained.

Chapter 4 presents two distinct pattern diversity architectures for mobile terminal applications. The first is a conformal wideband antenna with low SAR and orthogonal pattern diversity. The second is a gain compensated conformal shared ground design with high gain yield for minimal physical footprint. The proposed antenna is deployed in a real-world scenario and the received power profiles are presented for landscape and portrait modes.

Chapter 5 introduces the idea of path loss compensation in the context of indoor base station scenarios, followed by the designs of two distinct pattern diversity modules for base station applications. The antenna elements utilized in these modules are high aperture efficiency elements. The module is also investigated for a real-world scenario.

Chapter 6 introduces the principle of a shared aperture antenna followed by a design. A shared aperture antenna with orthogonal pattern diversity targeting base station applications is presented.

Chapter 7 includes a discussion on the need for co-design of 4G-LTE antennas and mmWave 5G pattern diversity modules. The miniaturization techniques for lower frequency bands is discussed, followed by a design example and a case study comparing the operation of 4G versus 5G in a real-world scenario.

Chapter 8 is dedicated to phased array designs at mmWave band. The primary design principles of phased array are discussed, followed by an exhaustive discussion of a design example of corner bent phased array design on a low-dielectric loss substrate.

Chapter 9 has a discussion on the nuances of fabrication and measurement challenges related to antenna designs beyond 20 GHz. A few points to be noted while performing antenna characterizations are also discussed.

Finally, Chapter 10 has an elaborate discussion on the upcoming research areas in the domain of millimeter wave antenna designs. Prospective designs and research gaps are also discussed in this chapter.

References

1. P. Rysavy. "Transition to 4G: 3GPP Broadband Evolution to IMT-Advanced (4G)." *3G Americas Publishes Research Report on 3GPP Mobile Broadband Evolution* (online), 2010.
2. Forecast, Global Mobile Data Traffic. "Cisco visual networking index: global mobile data traffic forecast update, 2017–2022." Update 2017 (2019): 2022.
3. C. Rowell and E. Y. Lam, "Mobile-phone antenna design," *IEEE Antennas and Propagation Magazine*, 54(4), 14–34, August 2012.
4. Ulf Sandell, Antenna device for portable equipment, United States Patent 5,661,495, August 26, 1997.
5. A. I. Sulyman, A. T. Nassar, M. K. Samimi, G. R. Maccartney, T. S. Rappaport, and A. Alsanie, "Radio propagation path loss models for 5G cellular networks in the 28 GHZ and 38 GHZ millimeter-wave bands," *IEEE Communications Magazine*, 52(9), 78–86, September 2014.
6. W. Roh et al., "Millimeter-wave beamforming as an enabling technology for 5G cellular communications: Theoretical feasibility and prototype results," *IEEE Communications Magazine*, 52(2), 106–113, February 2014.
7. T. S. Rappaport et al., "Millimeter wave mobile communications for 5G cellular: It will work!," *IEEE Access*, 1, 335–349, 2013.
8. G. Zhang et al., "Experimental characterization of millimeter-wave indoor propagation channels at 28 GHz," *IEEE Access*, 6, 76516–76526, 2018.

9. Y. Huo, X. Dong, and W. Xu, 5G cellular user equipment: From theory to practical hardware design," *IEEE Access*, 5, 13992–14010, 2017.
10. D. Psychoudakis, Z. Wang, and F. Aryanfar, "Dipole array for mm-wave mobile applications," *Proceedings of the IEEE Antennas Propagation Society International Symposium (APSURSI)*, Orlando, FL, USA, 660–661, July 2013.
11. I. Syrytsin, M. Shen, and G. F. Pedersen, "Antenna integrated with a microstrip filter for 5G Mm-wave applications," *2018 International Conference on Electromagnetics in Advanced Applications (ICEAA)*, Cartagena des Indias, 438–441, 2018.
12. S. X. Ta, H. Choo, and I. Park, Broadband printed-dipole antenna and its arrays for 5G applications, *IEEE Antennas and Wireless Propagation Letters*, 16, 2183–2186, 2017.
13. S. Zhang, X. Chen, I. Syrytsin, and G. F. Pedersen, "A planar switchable 3-D-coverage phased array antenna and its user effects for 28-GHz mobile terminal applications," *IEEE Transactions on Antennas Propagation*, 65(12), 6413–6421, December 2017.
14. N. Ojaroudiparchin, M. Shen, S. Zhang, and G. F. Pedersen, A switchable 3-D-coverage-phased array antenna package for 5G mobile terminals," *IEEE Antennas and Wireless Propagation Letters*, 15, 1747–1750, 2016.
15. N. Ojaroudiparchin, Ming Shen, and G. F. Pedersen, "Wide-scan phased array antenna fed by coax-to-microstriplines for 5G cell phones," *2016 21st International Conference on Microwave, Radar and Wireless Communications (MIKON)*, Krakow, 1–4, 2016.
16. I. Syrytsin, S. Zhang, and G. F. Pedersen, "User impact on phased and switch diversity arrays in 5G mobile terminals," *IEEE Access*, 6, 1616–1623, 2018.
17. J. Helander, K. Zhao, Z. Ying, and D. Sjöberg, "Performance analysis of millimeter-wave phased array antennas in cellular handsets," *IEEE Antenna Wireless Propagation Letters*, 15, 504–507, 2016.
18. K. Zhao, J. Helander, D. Sjoberg, S. He, T. Bolin, and Z. Ying, "User body effect on phased array in user equipment for 5G mm wave communication system," *IEEE Antenna Wireless Propagation Letters*, 16, 864–867, 2017.
19. S. Zhang, I. Syrytsin, and G. F. Pedersen, "Compact beam-steerable antenna array with two passive parasitic elements for 5G mobile terminals at 28 GHz," *IEEE Transactions on Antennas and Propagation*, 66(10), 5193–5203, October 2018.
20. I. Syrytsin, S. Zhang, G. F. Pedersen, and A. S. Morris, "Compact quad-mode planar phased array with wideband for 5G mobile terminals," *IEEE Transactions on Antennas and Propagation*, 66(9), 4648–4657, September 2018.
21. P. N. Choubey, W. Hong, Z. Hao, P. Chen, T. Duong, and J. Mei, "A wideband dual-mode SIW cavity-backed triangular-complimentary-split-ring-slot (TCSRS) antenna," *IEEE Transactions on Antennas and Propagation*, 64(6), 2541–2545, June 2016.
22. Q.-L. Yang, Y.-L. Ban, K. Kang, C.-Y.-D. Sim, and G. Wu, "SIW multibeam array for 5G mobile devices," *IEEE Access*, 4, 2788–2796, June 2016.
23. G. H. Zhai, W. Hong, K. Wu, and Z. Q. Kuai, "Wideband substrate integrated printed log-periodic dipole array antenna," *IET Microwaves, Antennas & Propagation*, 4(7), 899–905, July 2010.
24. W. El-Halwagy, R. Mirzavand, J. Melzer, M. Hossain, and P. Mousavi, "A substrate-integrated fan-beam dipole antenna with varied fence shape for mm-wave 5G wireless," *2018 IEEE International Symposium on Antennas and Propagation and USNC/URSI National Radio Science Meeting*, Boston, MA, USA, 251–252, 2018.
25. R. L. Haupt and M. Lanagan, "Reconfigurable antennas," *IEEE Antennas and Propagation Magazine*, 55(1), 49–61, February 2013.
26. J. T. Rayno and S. K. Sharma, "Wideband frequency-reconfigurable spirograph planar monopole antenna (SPMA) operating in the UHF band," *IEEE Antennas and Wireless Propagation Letters*, 11, 1537–1540, 2012.
27. R. K. Singh, A. Basu, and S. K. Koul, "A novel reconfigurable microstrip patch antenna with polarization agility in two switchable frequency bands," *IEEE Transactions on Antennas and Propagation*, 66(10), 5608–5613, October 2018.

28. R. K. Singh, A. Basu, and S. K. Koul, "A novel pattern-reconfigurable oscillating active integrated antenna," *IEEE Antennas and Wireless Propagation Letters*, 16, 3220–3223, 2017.
29. W. S. Chang, C. Yang, C. K. Chang, W. J. Liao, L. Cho, and W. S. Chen, "Pattern reconfigurable millimeter-wave antenna design for 5G handset applications," *2016 10th European Conference on Antennas and Propagation (EuCAP)*, Davos, 1–3, 2016.
30. S. Hage-Ali, N. Tiercelin, P. Coquet, R. Sauleau, H. Fujita, V. Preobrazhensky, and P. Pernod, "A millimeter-wave microstrip antenna array on ultra-flexible micromachined polydimethylsiloxane (PDMS) polymer," *IEEE Antennas Wireless Propagation Letter*, 8, 1306–1309, 2009.
31. M. Tang, T. Shi, and R. W. Ziolkowski, "Flexible efficient quasi-yagi printed uniplanar antenna," *IEEE Transactions on Antennas and Propagation*, 63(12), 5343–5350, December 2015.
32. A. Bisognin, J. Thielleux, W. Wei, D. Titz, F. Ferrero, P. Brachat, G. Jacquemod, H. Happy, and C. Luxey, "Inkjet coplanar square monopole on flexible substrate for 60-GHz applications," *Antennas and Wireless Propagation Letters IEEE*, 13, 435–438, 2014.
33. A. Rahimian, A. Alomainy, and Y. Alfadhl, "A flexible printed millimetre-wave beamforming network for WiGig and 5G wireless subsystems," *Antennas & Propagation Conference (LAPC) 2016*, Loughborough, 1–5, 2016.
34. S. F. Jilani, Q. H. Abbasi, and A. Alomainy, "Inkjet-printed millimetre-wave PET-based flexible antenna for 5G wireless applications," *5G Hardware and System Technologies (IMWS-5G) 2018 IEEE MTT-S International Microwave Workshop Series on*, 1–3, 2018.
35. S. F. Jilani, M. O. Munoz, Q. H. Abbasi, and A. Alomainy, "Millimeter-wave liquid crystal polymer based conformal antenna array for 5G applications," *Antennas and Wireless Propagation Letters IEEE*, 18(1), 84–88, 2019.
36. R. Bahr, B. Tehrani, and M. M. Tentzeris, "Exploring 3-D printing for new applications: Novel inkjet- and 3-D-printed millimeter-wave components interconnects and systems," *Microwave Magazine IEEE*, 19(1), 57–66, 2018.
37. B. S. Cook, B. Tehrani, J. R. Cooper, and M. M. Tentzeris, "Multilayer inkjet printing of millimeter-wave proximity-fed patch arrays on flexible substrates," *IEEE Antennas and Wireless Propagation Letters*, 12, 1351–1354, 2013.
38. A. Georgiadis, J. Kimionis, and M. M. Tentzeris, "3D/Inkjet-printed millimeter wave components and interconnects for communication and sensing," *2017 IEEE Compound Semiconductor Integrated Circuit Symposium (CSICS)*, Miami, FL, USA, 1–4, 2017.
39. S. F. Jilani and A. Alomainy, "Planar millimeter-wave antenna on low-cost flexible PET substrate for 5G applications," *2016 10th European Conference on Antennas and Propagation (EuCAP)*, Davos, 1–3, 2016.
40. B. Yu, K. Yang, C. Sim, and G. Yang, "A novel 28 GHz beam steering array for 5G mobile device with metallic casing application," *IEEE Transactions on Antennas and Propagation*, 66(1), 462–466, January 2018.
41. M. Stanley, Y. Huang, H. Wang, H. Zhou, A. Alieldin, and S. Joseph, "A capacitive coupled patch antenna array with high gain and wide coverage for 5G smartphone applications," *Access IEEE*, 6, 41942–41954, 2018.
42. B. Xu, Z. Ying, L. Scialacqua, A. Scannavini, L. J. Foged, T. Bolin, K. Zhao, S. He, and M. Gustafsson, "Radiation performance analysis of 28 GHz antennas integrated in 5G mobile terminal housing," *Access IEEE*, 6, 48088–48101, 2018.
43. J. Bang and J. Choi, "A SAR reduced mm-wave beam-steerable array antenna with dual-mode operation for fully metal-covered 5G cellular handsets," *IEEE Antennas and Wireless Propagation Letters*, 17(6), 1118–1122, June 2018.
44. R. Hussain, A. T. Alreshaid, S. K. Podilchak, and M. S. Sharawi, "Compact 4G MIMO antenna integrated with a 5G array for current and future mobile handsets," *IET Microwaves, Antennas & Propagation*, 11(2), pp. 271–279, 2017.
45. J. Kurvinen, H. Kähkönen, A. Lehtovuori, J. Ala-Laurinaho, and V. Viikari, "Co-designed mm-wave and LTE handset antennas," *IEEE Transactions on Antennas and Propagation*, 67(3), 1545–1553, March 2019.

46. S. Bodhisatwa, Y. T. JoakimHallin, S. Sahl, S. Reynolds, Ö. R. KristofferSjögren et al., "7.2 A 28GHz 32-element phased-array transceiver IC with concurrent dual polarized beams and 1.4 degree beam-steering resolution for 5G communication," *2017 IEEE International Solid-State Circuits Conference (ISSCC)*, San Francisto, CA, 128–129, 2017.
47. X. Gu, A. Valdes-Garcia, A. Natarajan, B. Sadhu, D. Liu, and S. K. Reynolds, "W-band scalable phased arrays for imaging and communications," *IEEE Communications Magazine*, 53(4), 196–204, April 2015.
48. B. Sadhu et al., "A 28-GHz 32-element TRX phased-array IC with concurrent dual-polarized operation and orthogonal phase and gain control for 5G communications," *IEEE Journal of Solid-State Circuits*, 52(12), 3373–3391, December 2017.
49. Y. Cheng and Y. Li, "A wideband high-gain quasi-yagi antenna for millimeter-wave applications," *2017 Sixth Asia-Pacific Conference on Antennas and Propagation (APCAP)*, Xi'an, 1–3, 2017.
50. R. A. Alhalabi and G. M. Rebeiz, "Differentially-fed millimeter-wave Yagi-Uda antennas with folded dipole feed," *IEEE Transactions on Antennas and Propagation*, 58(3), 966–969, March 2010.
51. Y. Ou and G. M. Rebeiz, "Differential microstrip and slot-ring antennas for millimeter-wave silicon systems," *IEEE Transactions on Antennas and Propagation*, 60(6), 2611–2619, June 2012.
52. O. M. Haraz, S. A. Alshebeili, and A. Sebak, "Low-cost high gain printed log-periodic dipole array antenna with dielectric lenses for V-band applications," *IET Microwaves, Antennas & Propagation*, 9(6), pp. 541-552, 24 4, 2015.
53. G. Zhai, Y. Cheng, Q. Yin, S. Zhu, and J. Gao, "Gain enhancement of printed log-periodic dipole array antenna using director cell," *IEEE Transactions on Antennas and Propagation*, 62(11), 5915–5919, November 2014.
54. G. Zhai, Y. Cheng, Q. Yin, S. Zhu, and J. Gao, "Uniplanar millimeter-wave log-periodic dipole array antenna fed by coplanar waveguide," *International Journal of Antennas and Propagation* 2013.
55. J. Park, J. Ko, H. Kwon, B. Kang, B. Park, and D. Kim, "A tilted combined beam antenna for 5G communications using a 28-GHz band," *IEEE Antennas and Wireless Propagation Letters*, 15, 1685–1688, 2016.
56. O. M. Haraz, A. Elboushi, S. A. Alshebeili, and A. Sebak, "Dense dielectric patch array antenna with improved radiation characteristics using ebg ground structure and dielectric superstrate for future 5G cellular networks," *IEEE Access*, 2, 909–913, 2014.
57. Y. Li and K. M. Luk, "Wideband perforated dense dielectric patch antenna array for millimeter-wave applications," *Antennas and Propagation IEEE Transactions on*, 63(8), 3780–3786, 2015.
58. J. Xu, W. Hong, H. Tang, Z. Kuai, and K. Wu, "Half-mode substrate integrated waveguide (HMSIW) leaky-wave antenna for millimeter-wave applications, *IEEE Antennas and Wireless Propagation Letters*, 7, 85–88, 2008.
59. X. Bai, S. Qu, C. H. Chan, and K. B. Ng, "Millimeter-wave leaky-wave antenna based on dielectric filled metal groove waveguide," *2015 IEEE 4th Asia-Pacific Conference on Antennas and Propagation (APCAP)*, Kuta, 370–372, 2015.
60. M. Khalily, R. Tafazolli, T. A. Rahman, and M. R. Kamarudin, "Design of phased arrays of series-fed patch antennas with reduced number of the controllers for 28-GHz mm-wave applications," *Antennas and Wireless Propagation Letters IEEE*, 15, 1305–1308, 2016.
61. M. Khalily, R. Tafazolli, P. Xiao, and A. A. Kishk, "Broadband mm-wave microstrip array antenna with improved radiation characteristics for different 5G applications," *Antennas and Propagation IEEE Transactions on*, 66(9), 4641–4647, 2018.
62. C. X. Mao, M. Khalily, P. Xiao, T. W. C. Brown, and S. Gao, "Planar sub-millimeter-wave array antenna with enhanced gain and reduced sidelobes for 5G broadcast applications," *Antennas and Propagation IEEE Transactions on*, 67(1), 160–168, 2019.
63. A. Edalati and T. A. Denidni, "High-gain reconfigurable sectoral antenna using an active cylindrical FSS structure," *IEEE Transactions on Antennas and Propagation*, 59(7), 2464–2472, July 2011.
64. A. Edalati and T. A. Denidni, "Frequency selective surfaces for beam-switching applications," *IEEE Transactions on Antennas and Propagation*, 61(1), 195–200, January 2013.

65. J. Li, Q. Zeng, R. Liu, and T. A. Denidni, "A compact dual-band beam-sweeping antenna based on active frequency selective surfaces," *Antennas and Propagation IEEE Transactions on*, 65(4), 1542–1549, 2017.
66. J. Li, T. A. Denidni, and Q. Zeng, "A dual-band reconfigurable radiation pattern antenna based on active frequency selective surfaces," *Antennas and Propagation (APSURSI) 2016 IEEE International Symposium on*, 1245–1246, 2016.
67. S. M. Mahmood and T. A. Denidni, "Pattern-reconfigurable antenna using a switchable frequency selective surface with improved bandwidth," *Antennas and Wireless Propagation Letters IEEE*, 15, 1148–1151, 2016.
68. H. Aliakbari, A. Abdipour, R. Mirzavand, A. Costanzo, and P. Mousavi, "A single feed dual-band circularly polarized millimeter-wave antenna for 5G communication," *2016 10th European Conference on Antennas and Propagation (EuCAP)*, Davos, 1–5, 2016.
69. M. Nosrati and N. Tavassolian, "A single feed dual-band, linearly/circularly polarized cross-slot millimeter-wave antenna for future 5G networks," *2017 IEEE International Symposium on Antennas and Propagation & USNC/URSI National Radio Science Meeting*, San Diego, CA, 2467–2468, 2017.
70. S. Park and S. Park, "LHCP and RHCP substrate integrated waveguide antenna arrays for millimeter-wave applications," *IEEE Antennas and Wireless Propagation Letters*, 16, 601–604, 2017.
71. H. Al-Saedi, W. M. Abdel-Wahab, S. Gigoyan, R. Mittra, and S. Safavi-Naeini, "Ka-band antenna with high circular polarization purity and wide AR beamwidth," *Antennas and Wireless Propagation Letters IEEE*, 17(9), 1697–1701, 2018.
72. E. Al Abbas, A. T. Mobashsher, and A. Abbosh, "Polarization reconfigurable antenna for 5G cellular networks operating at millimeter waves," *2017 IEEE Asia Pacific Microwave Conference (APMC)*, Kuala Lumpur, Malaysia, 772–774, 2017.
73. Z. Gan, Z. H. Tu, Z. M. Xie, Q. X. Chu, and Y. Yao, "Compact wideband circularly polarized microstrip antenna array for 45 GHz application," *Antennas and Propagation IEEE Transactions on*, 66(11), 6388–6392, 2018.
74. M. Asaadi and A. Sebak, "High-gain low-profile circularly polarized slotted SIW cavity antenna for MMW applications," *IEEE Antennas and Wireless Propagation Letters*, 16, 752–755, 2017.
75. Y. Li, Z. N. Chen, X. Qing, Z. Zhang, and J. X. Z. Feng, "Axial ratio bandwidth enhancement of 60-GHz substrate integrated waveguide-fed circularly polarized LTCC antenna array," *IEEE Transactions on Antennas Propagation*, 60(10), 4619–4626, October 2012.
76. C. Liu, Y. X. Guo, X. Bao, and S. Q. Xiao, "60-GHz LTCC integrated circularly polarized helical antenna array," *IEEE Transactions on Antennas Propagation*, 60(3), 1329–1335, March 2012.
77. A. Dadgarpour, M. Sharifi Sorkherizi, and A. A. Kishk, "Wideband low-loss magnetoelectric dipole antenna for 5G wireless network with gain enhancement using meta lens and gap waveguide technology feeding," *IEEE Transactions on Antennas and Propagation*, 64(12), 5094–5101, December 2016.
78. M. Sharifi Sorkherizi, A. Dadgarpour, and A. A. Kishk, "Planar high-efficiency antenna array using new printed ridge gap waveguide technology," *IEEE Transactions on Antennas and Propagation*, 65(7), 3772–3776, July 2017.
79. P. A. Dzagbletey, K. Kim, W. Byun, and Y. Jung, "Stacked microstrip linear array with highly suppressed side-lobe levels and wide bandwidth," *IET Microwaves, Antennas & Propagation*, 11(1), pp. 17-22, 8 1, 2017.
80. S. Afoakwa an dY. B. Jung, "Wideband microstrip comb-line linear array antenna using stubbed-element technique for high sidelobe suppression," *Antennas and Propagation IEEE Transactions on*, 65(10), 5190–5199, 2017.
81. B. Yang, Z. Yu, Y. Dong, J. Zhou, and W. Hong, "Compact tapered slot antenna array for 5G millimeter-wave massive MIMO systems," *IEEE Transactions on Antennas and Propagation*, 65(12), 6721–6727, December 2017.

82. S. Zhu, H. Liu, Z. Chen, and P. Wen, "A compact gain-enhanced vivaldi antenna array with suppressed mutual coupling for 5G mm wave application, *Antennas and Wireless Propagation Letters IEEE*, 17(5), 776–779, 2018.
83. T. Djerafi and K. Wu, "Corrugated substrate integrated waveguide (SIW) antipodal linearly tapered slot antenna array fed by quasi-triangular power divider," *Progress in Electromagnetic Research C*, 26, 139–151, 2012.
84. A. Ghiotto, F. Parment, T.-P. Vuong, and K. Wu, "Millimeter-wave air-filled SIW antipodal linearly tapered slot antenna," *IEEE Antennas Wireless Propagation Letters*, 16, 768–771, 2016.
85. E. B. Ayman and R. Sarkis, "Design of tilted taper slot antenna for 5G base station antenna circular array," *Antennas and Propagation (MECAP), 2016 IEEE Middle East Conference on IEEE*, 1–4, 2016.
86. L. Ranzani, D. Kuester, K. J. Vanhille, A. Boryssenko, E. Grossman, and Z. Popović, "G-band micro-fabricated frequency-steered arrays with 2/GHz beam steering, *IEEE Transactions on Terahertz Science and Technology*, 3(5), 566–573, September 2013.

2

Conformal Antennas for Mobile Terminals

2.1 Introduction

The typical requirements for antenna designs for millimeter wave 5G mobile terminals are introduced in this chapter. Conventionally, antennas designed for mobile terminals must be future proof and be compliant with various sub-bands of operation of the cellular network. The antenna must also operate independently of the network provider, hence wideband antennae are preferable in the context of a mobile terminal. However, wideband antennae do not necessarily mean that high gain could be achieved for the available aperture of the antenna. A simple Vivaldi based antenna design would be an obvious solution to both wideband and high gain, but the physical footprint would be compromised.

Hence, wideband high gain conformal antenna elements are presented in this chapter targeting mobile terminal applications. The first design is a CPW-fed wideband slot antenna. As the coplanar waveguide feeding technique is uncommon in Ka band designs, a thorough explanation regarding this is presented. The antenna is excited by a stepped impedance transformer to a slot radiator with a wide impedance bandwidth. The proposed antenna is investigated for a 90° bend, but since the slot radiates in both the directions because of the orthogonal parasitic ground, an electrically close reflector with linear phase characteristics is designed and integrated behind the radiator to increase the forward gain of the conformal antenna, thus leading to a wideband, high gain and compact antenna. The presented antenna could easily be integrated with commercial smartphones. The second design has the same radiating element with CPW feeding and corner bending backed by an exponentially tapered reflector to yield high gain for minimal effective radiating volume. In this design, a 3D-printed scaffolding has been designed, over which a copper film is pasted to act as an exponentially tapered reflector. This design is electrically compact compared to reported designs, in addition to yielding a high gain in the forward direction.

2.2 Typical Requirements for Mobile Antennas

The desirable features of mmWave 5G mobile terminal antennas include:

- *Compact design* With the least possible physical footprint for the designed gain. Antennas conformal to the smartphone structure would be desirable, since the RF real estate available is scarce. Typical dimensions of the smartphone panel is 10 cm × 8 cm × 0.5 cm. The proposed antennas must fit at the edge of these panel dimensions with a minimal metallic footprint. It is well known that an increase in the frequency of operation would lead to a corresponding decrease in the wavelength, consequently leading to a decrease in the physical size of the antenna to achieve similar characteristics to the antenna. The frequency of 28 GHz has a half-wavelength of around 5 mm, which is the depth of typical smartphones available on the market today. Hence single elements could be accommodated without tweaking around with the panel depth, but beam scanning or pattern diversity modules might lead to an increase in the physical footprint. Therefore there is a trade-off between the available physical footprint within the constraints of the smartphone, gain to be achieved for the desirable link budget for a given scenario, and the desired angular coverage to avoid signal blockage.

- *Reasonable SAR* The SAR values must be below the prescribed limit to channel radiation towards the base station and away from the user in multi-orientation modes. Historically, SAR reduction has been a very important design constraint, since the sub-6 GHz has higher penetration into human tissue and might lead to unwanted heating effects. In contrast to the previous wireless standards, the SAR requirements are not very stringent, but it is better to design antennas with the least possible radiation towards the user. The heating effects of millimeter waves is negligible, but human tissue acts as a strong attenuator when illuminated by a millimeter wave signal, therefore low SAR is recommended to maintain the link budget with the base station.

- *High gain, narrow beam* High gain antennas (>10 dBi) would be desirable to design the communication link with a feasible link budget. The gain must be maximized for the available radiating aperture of the mobile phone mount. It must also be noted that a high gain would mean a narrow beam. In order to maintain the data link, a beam steering mechanism might be desirable. Higher gain might easily be designed by increasing the electrical aperture of the antenna, but the antenna would also suffer from poor coverage because of the decreased beamwidth. As a general rule of thumb, the antennas on the mobile terminal must have a higher beamwidth compared to their counterparts at the base station. Ideally, a hemispherical pattern in one of the principal planes with high front-to-back ratio and minimal radiation towards the user is recommend.

- *Quick switching* In order to maintain a jitter-free data link with mobility of the user, the narrow beams must be switched within a few μs. The antenna system must be distortion-free and must not emit spurious radiation during the switching. The isolation between the antenna system and the RF front end must be high (>30 dB). If a phased array system is implemented, the beam scanning must be quick enough without any compromise in the received power, and the received power must be independent of the user's mobility. Also, if pattern diversity or polarization diversity modules are implemented, the mutual coupling among the elements must also be minimal.

- *Co-existence with 4G* The current commercial smartphone is pretty crowded inside its panel, with the RF electronics, wireless chipsets, switches and its corresponding antennas. Since most of the telecommunication products feature backward compatibility, mmWave antennas operating in the 28 GHz band must have minimal mutual coupling with the existing 4G LTE MIMO antennas operating in the 0.7–2.7 GHz band. Hence 4G antennas must also be re-designed with integrated harmonic filters to prevent multi-octave coupling with the 5G antennas.

Conformal Antennas for Mobile Terminals

- *Low cost* If industry-standard processes were to be utilized for manufacturing antennas and their feed system in the mmWave band, the production cost per unit could be reduced. Ideally, standard manufacturing techniques already in existence for 4G must also be used for manufacturing 5G antennas. Furthermore, the resolution or minimum dimension of the feature size of the designed antennas must be compliant with the industry-standard processes.

2.3 CPW-fed Wideband Corner Bent Antenna for 5G Mobile Terminals

Antennas could be classified generally into three types, namely broadside radiators, endfire radiators, and omnidirectional radiators, as reported in [1–18]. Omnidirectional antennas have very low gain and hence are unsuited for mmWave 5G applications, be they on the mobile terminal or at the base station. The use case in this section is the antenna to be integrated with the mobile terminal. In this chapter, broadside antennas are investigated and their applications in integrating with a typical smartphones is illustrated. Imagine a broadside radiator mounted on a mobile platform. Researchers believe that mmWave 5G would be used primarily for data transmission rather than voice, as voice services would be catered for by previous wireless standards. In this context, the antennas to be designed must be compliant with this use case. The planar broadside antenna would radiate towards the user when the user is engaged in data usage mode, as illustrated in Figure 2.1.

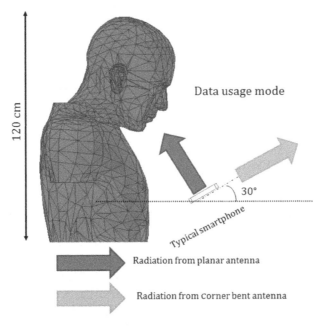

FIGURE 2.1
Comparison of planar versus corner bent antenna integrated on a mobile phone.

The assumption is that the antenna has a hemispherical beam radiating in the boresight orientation. With broadside radiators, the radiation is towards the user, and the human head or torso increases the signal attenuation towards the base station. Conventional broadside radiators as they are might not be a clever solution for mobile terminal antennas. End-fire antennas with high front-to-back ratio and forward-directed beam might be a possible solution to this problem, where blockage by the human body is completely avoided. But the physical footprint of the antenna would be high, consuming valuable real estate in the smartphone, even though the antenna is planar RF circuitry and the motherboard would be integrated with the antenna, hence increasing the mutual coupling and interference between the antenna and the RF motherboard. Therefore, a corner bent antenna with high gain and bandwidth is presented in this section. It must be noted that the antenna to be designed must radiate minimally towards the user and have an isolating network when integrated with the motherboard. The reflector in the design would be acting as a natural isolating network for avoiding interference with the motherboard.

Various configurations of antennas for 5G mobile terminals have been investigated, such as [7], which has multi-layered design, hence occupying a relatively larger area, with a 9% impedance bandwidth. Even though the antenna element proposed in [8] has a 10% bandwidth, the radiation patterns might not be suitable for integration with the mobile terminal. The element proposed in [9] has a bandwidth of more than 35% with an end fire gain of 4–6 dBi, but the design is planar. The design presented in [10] offers 20% bandwidth with a SIW feed, hence increasing the complexity of the design. It must be noted that most of the reported designs are planar and designed with microstrip feed.

Conformal designs have been proposed earlier, such as in [11], meant for lightweight UAV applications. This design process could be adapted for a mobile terminal, but the scaffolding of the antenna would contribute to losses in addition to impedance mismatch. The SIW conformal antenna proposed in [12] would increase the complexity of the fabrication process. The conformal antenna proposed in [13] is designed at 60 GHz and mounted on a cylindrical surface with a radius of 25 mm, hindering its utility in 5G mobile terminals at 28 GHz. CPW feed would be the preferred method of feeding conformal antennas to be integrated with mobile terminals, to facilitate uniplanar transition.

CPW feed is feasible at lower frequencies up to 15 GHz, as demonstrated in [14], as the dimensions of the 50 Ω feed line would permit the application of a standard, inexpensive chemical etching process for fabrication. The CPW feed is also popular at mmWave frequencies and beyond, since the dimensions of the transmission line would be readily feasible with micro-fabrication technologies [15].

The CPW-fed LPDA antenna proposed in [16] has a gain of 1.5–4 dBi and is planar in nature. Even though the CPW-fed mmWave antenna demonstrated in [17] has an operating frequency of 30 GHz, the bandwidth is low (less than 5%) and it uses an air bridge in the feed line for impedance match. Hence a wideband CPW-fed corner bent antenna is presented.

2.3.1 CPW-fed Wideband Antenna

The proposed CPW-fed slot antenna is illustrated in Figure 2.2(a); all the dimensions are in millimetres. It is constructed on a Nelco NY9220 substrate with ε_r of 2.2 and 0.508 mm thickness. Typically, substrates are characterized for the dielectric constant at 10 GHz, and the manufacturer would specify the tolerance for the substrate. In this case, the dielectric constant varies around 2.2 ± 0.02. It would be good design practice to measure the

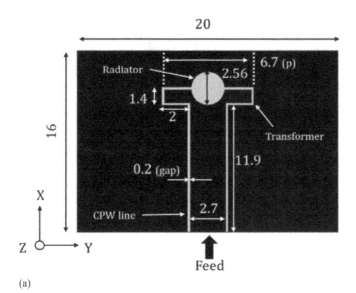

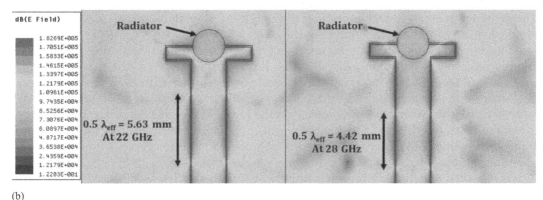

FIGURE 2.2
(a) CPW-fed wideband slot antenna design (all dimensions are in mm). (b) E-field plot at 22 and 28 GHz [23].

dielectric constant of the available substrate at the desired frequency by making a ring resonator or a standard inset-fed patch antenna. The loss tangent also also be calculated by measuring the transmission loss for a standard 50 Ω line over a 50 mm line of the substrate using a properly calibrated vector network analyser. It is also recommended to measure the thickness of the substrate by using digital Vernier calipers or a digital screw gauge. The thickness of copper would be around 15–35 μm. The thickness of the copper would not affect the performance of the antenna, but would decide the power handling capacity of the antenna. Ultimately, the antenna would be integrated within a mobile phone whose maximum transmitted power would be +30 dBm, and the transmissions would be in bursts. In this context, using thin substrates would be operational even after a longer usage. Since the substrate is only 0.5 mm thick, the inhomogeneity of the dielectric constant would be minimal compared to 30 mil FR4 substrates at the lower frequency. It is a common observation that the measured lower frequency input reflection coefficient measurements usually suffer from detuning primarily as a result of this effect.

The antenna could have been also designed on the popular Rogers 5880 substrate, but as it is a demonstration of the principle of corner bending it has been designed on Nelco NY9220. The low frequency substrates such as FR4 would perform poorly at high frequencies because of a high dielectric constant and its corresponding dielectric loss tangent, which leads to gain deterioration. A low dielectric constant is chosen to keep the surface wave modes to a minimum as opposed to a higher dielectric constant substrate. Electrically thin substrate was chosen to keep the cross-pol radiation to a minimum in the end fire. A 2.7 mm wide CPW line with a gap of 0.2 mm is chosen for the feed line, since wider trace lines larger than half a wavelength at the frequency of interest would lead to an over-moded antenna and hence result in dual-beam with poor patterns in the broadside. Also, when wider feed lines are used for feeding the antenna structure, standing waves might be created on the feedline, leading to poor impedance mismatch. The wider lines might also lead to radiation in unnecessary orientations. The circular slot is the primary radiator in this topology. The feed is a 59 Ω CPW line in series with a stepped impedance transformer of 50 Ω followed by the capacitive slot, which has a high impedance of around 80 Ω leading to an input impedance of around 65 Ω, hence achieving $|S_{11}|$ less than −10 dB. Figure 2.2(b) illustrates the E-field patterns at 22 and 28 GHz of the proposed antenna; the half-wavelength transmission line mode is also illustrated. The designer must consider that not all of the E-fields depicted in these kinds of graphs would lead to radiation. By analysing the co-pol and cross-pol radiation patterns, one must understand which part of the antenna creates the dominant radiating fields. It is a common misconception that only 50 Ω feed lines must be used for the design of transmission line feeding the antenna. This is not necessarily true; the objective of a good antenna design is to achieve an input reflection coefficient of less than −10 dB. This could happen as long as the input impedance is in the range of 27–95 Ω. Hence a 50 Ω feed line is not really required to get the desired input impedance bandwidth as the load to the transmission line could be varied by adjusting the transformer geometry and its corresponding dimensions.

It is observed that the circular slot is uniformly illuminated irrespective of the frequency of operation, hence a wide impedance bandwidth is achieved. The equivalent circuit is shown in Figure 2.3(a) and this was simulated in Agilent ADS. The concept of an equivalent circuit gives an estimate of antenna behaviour in the circuit domain, but gives an inadequate representation of the antenna model. However, this model aids in understanding the impedance behaviour of the antenna. The aperture and patterns of the antenna cannot be understood with this method. The diameter of the radiating slot matches a quarter-wavelength at 28 GHz, but it also supports a finite bandwidth to aid in radiation. An impedance match at the lower end of the frequency range could be achieved with a diameter tuning near 2 mm, for the aforementioned reason.

The distance between the feed and the radiating aperture is maintained at more than 1λ to reduce mutual coupling between the end-launch connector and the antenna. This is standard design practice in the industry, and for the measurement setup used in this section is a mandatory requirement. If the measurement setup is with a rotating horn antenna and the test antenna mounted on the test jig, then the electrically long antenna might not be required in the first place. The overall width of the antenna is sufficient for the electrically large end-launch connector (2.92 mm Southwest model) mount and to maintain broadside patterns. Note that this design constraint is followed for all the designs presented in this book. The T-shaped transformer is key for achieving the impedance match over more than 30% of the bandwidth. The connector holes must be appropriately designed

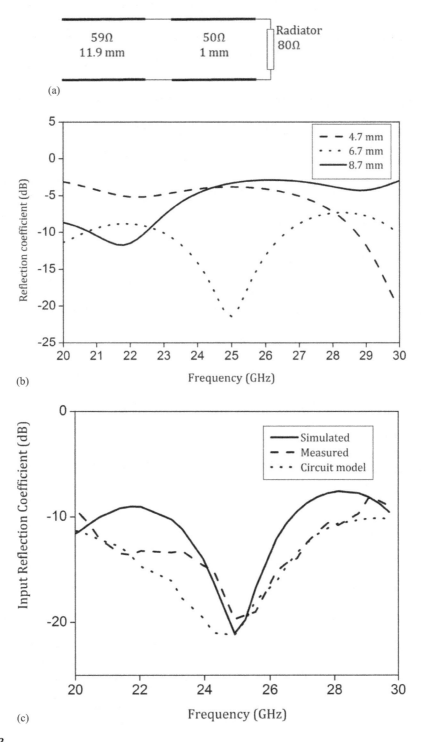

FIGURE 2.3
(a) Equivalent circuit of the proposed element; (b) Reflection coefficient variation with p; (c) Reflection coefficient of the proposed CPW-fed antenna [23].

in the mask during fabrication. These connector holes must be drilled in the substrate for mounting the antenna. The connection is primarily to align the antenna with the trace of the end-launch connector and tighten the screws.

The points of contact of the antenna's trace and the ground must be scraped adequately to ensure proper contact with the connector. The connector manufacturer suggests that the antenna or any other device must be soldered with the trace of the connector. The problem with this method is that the trace of the connector, which is barely 100 µm, has to be soldered. This requires precision soldering and a careful soldering technique. This might not be a feasible solution as the connector is expensive compared to its lower frequency counterpart.

Also, soldering might damage the sensitive trace of the connector. For all the aforementioned reasons, only solder-free transitions were used in all examples throughout the book. A solder-free transition would also permit the user to re-use the connector multiple times. The only consideration the designer must take into account is that the connector must be fastened properly – too tight tightening might snap the trace pin of the connector, and loose screwing would not lead to a proper energy transfer from the connector to the antenna or any device under test.

The input reflection coefficient variation with width of the stepped impedance transformer (p) is shown in Figure 2.3(b). For the given constraints on the feed line and the radiator, the impedance transformer action is studied to gain an insight into the transformer action. For the chosen quarter-wave aperture dimension and the CPW feed line, the optimal width of the transformer was found to be 6.7 mm, which is evident from the curves. The contour of the slot is equally important to establish the impedance match and beam integrity. For example, square or elliptical slots would result in less than 15% bandwidth because of improper mode excitations. The slot does not have a ground plane beneath it, therefore the radiation would be on both the sides of the antenna, leading to bi-directional beams.

Since the radiating aperture is not backed by a ground plane, a symmetric beam is formed on both sides of the aperture. The simulated and measured input reflection coefficient is depicted in Figure 2.3(c). All the full-wave simulations were performed in the industry standard Ansys HFSS. A perfect electric conductor bridge was used for CPW feeding in the simulation. A test simulation with the connector design was also performed to illustrate the effects of the connector on the antenna. The reflection coefficient was almost invariant with and without the connector, but the simulation time increased phenomenally because of the composite structure of electrically large metal with the reasonable electrical size of the antenna. Benchmarking of the connectors was done previously to understand the impedance of the connectors. The measured impedance bandwidth is from 20 to 28 GHz (34%). The variation in input reflection coefficient is minimal even with a gap of up to 0.3 mm in the CPW feedline, which indicates that the design is resilient to tolerance in fabrication errors – a more detailed analysis regarding flaws in the fabrication is carried out in Chapter 9. The measurements were done using Agilent PNA E8364C. The discrepancy could be attributed to fabrication tolerances, and the deviation between port impedance used in simulation and the impedance offered by the end-launch connector. A multi-segmented impedance transformer would have resulted in a higher bandwidth, with a re-designed profile of the radiator, but the broadside patterns would be more specular.

The radiation patterns in both the principal planes, i.e., E-plane (XZ plane) and H-plane (YZ plane) at 22, 25 and 28 GHz are shown in Figure 2.4. The front-to-back ratio is almost

Conformal Antennas for Mobile Terminals

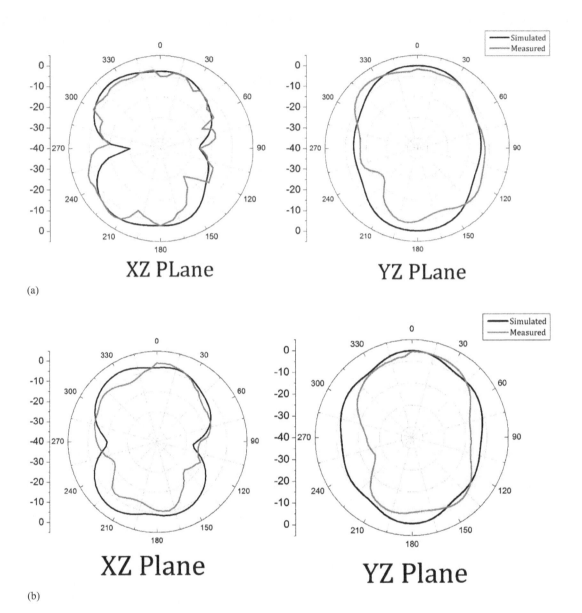

FIGURE 2.4
Patterns in XZ and YZ planes at: (a) 22 GHz and (b) 28 GHz [23].

0 dB. The deviations between simulated and measured patterns are caused by alignment errors and adapters utilized for measurements in the anechoic chamber.

The broadside gain is illustrated in Figure 2.5. The simulated gain varies from 2 to 3 dBi across the bandwidth. Since the measured beam widths in the H-plane (YZ) are more than 50° across the band, a low but consistent gain has been observed. Gain measurements were performed with the standard gain transfer method using Keysight horn antennas. It must be noted that to mitigate the path loss at 28 GHz, high gain antennas must be incorporated in the mobile terminals.

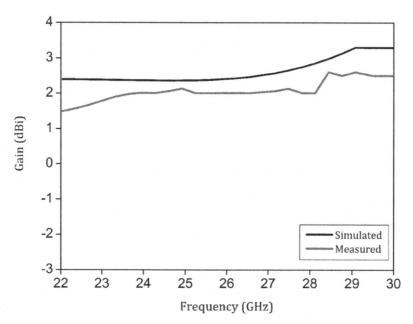

FIGURE 2.5
Broadside gain of the proposed antenna [23].

2.3.2 CPW-fed Corner Bent Antenna

The proposed antenna occupies an electrically large volume (1.6λ × 2 λ), hence it might be unsuitable for mobile terminals as it stands and therefore a corner bent architecture is proposed. The CPW-fed slot antenna is bent in front of the stepped impedance transformer, to create a uniform illumination towards the base station, but because of the lack of a ground plane, the beam is split towards and away from the base station. The choice of 20 mil substrate is justified because of its structural stability even after the introduction of the discontinuity. An ideal choice for corner bent architecture would be 5–10 mil substrates, but the dielectric scaffolding on which the antenna was supported would create an additional detuning of the antenna. Substrates above 30 mil would be much more stable structurally, but the antenna would have to be built in a piecewise manner, hence creating a substantial discontinuity in the feed line, which would lead to significant distortions in the radiation patterns. This strategy of building a piecewise antenna is a common practice in the lower frequencies, but it would create a heavier leakage of field lines leading to poor energy flow from the port to the radiator. The bending itself might lead to radiation if the field lines happened to be in-phase. This scenario might occur in some areas of the discontinuity.

It must also be noted that a uniplanar feed would create the least discontinuity when the radiating structure is conformed on to the mobile terminal. Conventional microstrip feed would lead to increased discontinuity in the ground plane, as the 17 μm of copper on the ground plane is exposed to more strain, hence leading to a higher fracture of the copper trace and in turn to a poor transition to the radiator.

The corner bent antenna design is illustrated in Figure 2.6. The radiating aperture was maintained with a clearance away from the orthogonal ground to create a reasonable beam on both sides in the XY plane. The distance between the feed and the aperture is still at 1λ to facilitate good radiation pattern measurements with the end launch connector. It is

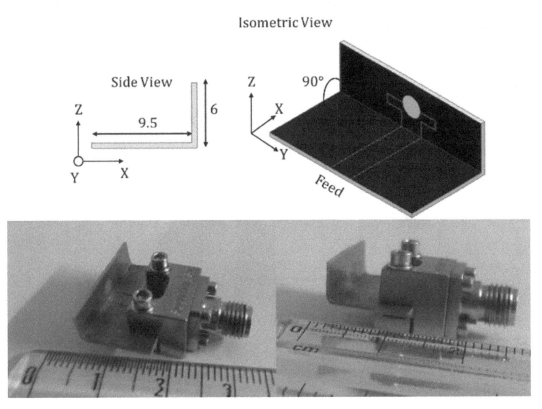

FIGURE 2.6
CPW fed corner bent antenna with photographs [23].

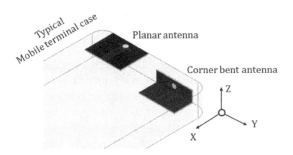

FIGURE 2.7
Comparison of antennas on a mobile terminal [23].

evident from Figure 2.7 that the radiation from a conventional antenna would be directed towards the user when in use. The proposed topology occupies a lesser volume compared to the planar design.

The simulated and measured input reflection coefficient of the corner bent antenna is shown in Figure 2.8. The measured impedance bandwidth is from 28.5 to 33.5 GHz, which is 16.2%. The reduction in impedance bandwidth is caused by the introduction of discontinuity. This discontinuity cannot be properly modelled in circuit theory to show its effect on the input impedance bandwidth. The discrepancy between simulated and measured

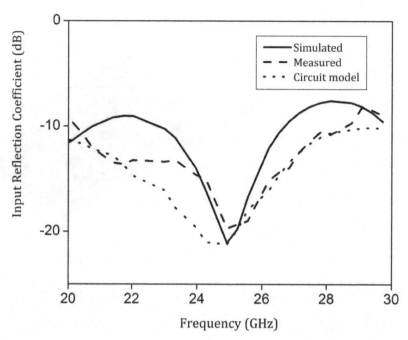

FIGURE 2.8
Reflection coefficient of corner bent antenna [23].

curves is caused by the non-ideal bending of the 20-mil copper-clad dielectric. Also, the corner bent model was assumed to incorporate a perfect 90° bend in simulation, but a minor alignment error could have crept in during the implementation of the antenna. The reflection coefficient deviation from the planar and corner bent designs could be reduced with a CPW-fed LPDA (log periodic dipole antenna), but achieving uniform gain for more than 20% bandwidth might be a challenging task. To maintain the input impedance behaviour of the planar and its corner bent counterpart, an additional impedance transformer could be designed, but since the reflection coefficient is within acceptable limits for the frequency of interest, an additional design tweak was not necessary. The inductive corner bent discontinuity creates an input impedance of (42 − j11) Ω and (61 + j7) Ω at 27 GHz and 31 GHz, respectively, hence creating a dual mode of operation, but the $|S_{11}| < -10$ dB is achieved for a wider band.

The E-plane (XZ plane) co-pol and cross-pol patterns are shown in Figure 2.9, from 27 to 30 GHz, in steps of 1 GHz. The measured front beamwidth is around 25° ± 10° and the front-to-back ratio is 0.5 dB in the E-plane. The front half-space is primarily caused by the radiating slot, and the radiation on the other side is because of the scattering effects from the electrically large ground in the orthogonal direction. It must also be observed that the effective aperture has been increased in the E-plane as a result of the parasitic ground, resulting in a decrease of beamwidth compared to the planar design, consequently increasing the gain by almost 2 dB in the bandwidth of interest. The H-plane (XY plane) patterns are shown in Figure 2.10. Since the parasitic ground plane is in the orthogonal plane, it has minimal influence on the H-plane patterns. The front beamwidth is around 60° ± 10°, with a front-to-back ratio of 1 dB, illustrating that the beam is split almost identically, with slightly higher radiation in the forward direction compared to the radiation towards the user, when mounted on a mobile terminal.

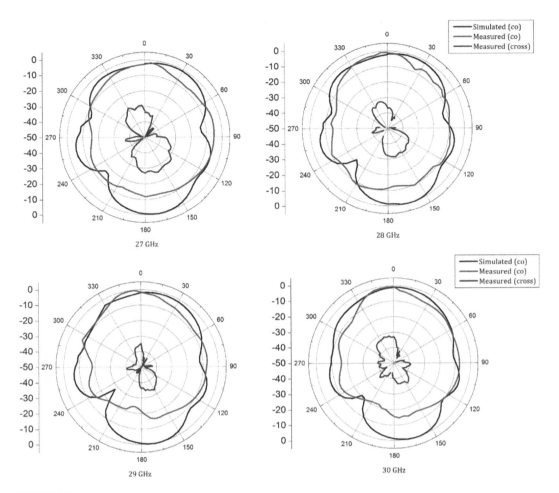

FIGURE 2.9
Patterns in XZ plane in the frequency band 27–30 GHz [23].

The measured radiation patterns have a deviation in the back-lobe due to utilization of the electrically large adapter for pattern measurements. The forward gain in the XY plane varies from 4 to 5.6 dBi in the frequency span from 2 to 33 GHz. Measured gain is also depicted in Figure 2.11. Since, the gain is stable across 27.5% of the bandwidth, the proposed corner bent antenna could be used for 5G mobile terminals. The radiation patterns of the corner bent antenna illustrate that energy is directed towards the base station and the user. The proposed topology would also lead to higher specific absorption rate. In order to mitigate this effect, the back lobe must be diminished and gain enhanced towards base station.

Several techniques could be incorporated for reduction of the back lobe: a localized ground plane could be designed which would essentially be effective for a narrowband. An absorber such as a Salisbury screen could be placed a quarter-wavelength away from the radiating aperture, but there the gain would be reduced. The third technique is to place a wideband reflector for back-lobe reduction leading to gain enhancement in the forward direction. Since this architecture would provide relatively higher gain for a wider band compared to other methods, the reflector design is investigated.

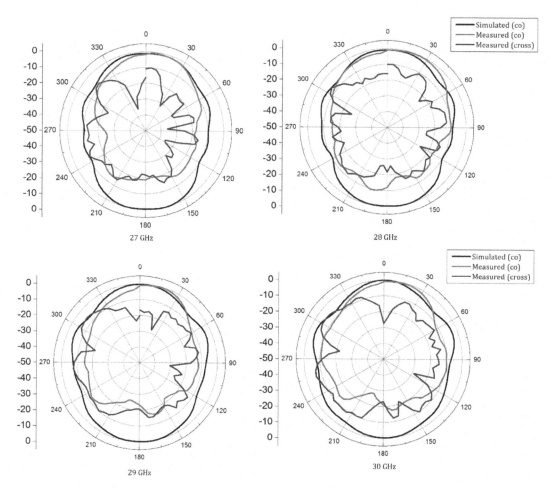

FIGURE 2.10
Patterns in XY plane in the frequency band 27–30 GHz [23].

2.3.3 CPW-fed Corner Bent Antenna with Reflector

The obvious solution for the design of a reflector would be to mount an electrically large (at least $3\lambda \times 3\lambda$) piece of metal behind the radiating aperture at a quarter-wavelength of the centre frequency, but this topology would decrease the impedance bandwidth, therefore a wideband reflector with periodic structure is proposed. It must also be noted that if the reflector size is 3λ, i.e., 30 mm at 28 GHz, this dimension would be unacceptable, since the panel height of typical smartphones is 10–15 mm. Hence the antenna integrated with a reflector must offer high impedance bandwidth, gain and compact size. A linear phase structure is essential for gain enhancement, therefore the slots are strategically etched, as shown in the design evolution schematics of Figure 2.12(a), and the corresponding phase variation against frequency is illustrated in Figure 2.12(b). The proposed unit cell and the simulation model are shown in Figure 2.13(a). Conventional cross slots would be narrowband, and the other alternative is to design multi-layered reflectors to create a net effect of wideband reflection; this topology would increase the

Conformal Antennas for Mobile Terminals

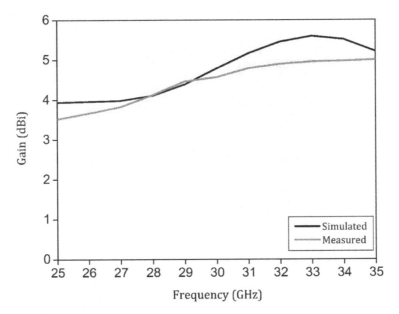

FIGURE 2.11
Forward gain of the corner bent antenna [23].

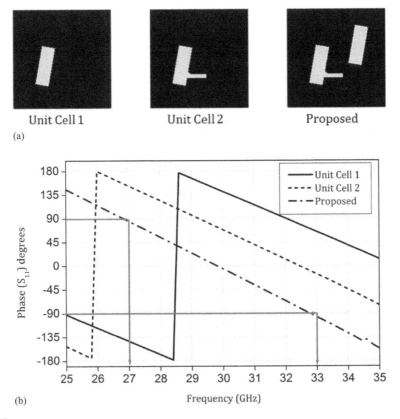

FIGURE 2.12
(a) Design evolution of the unit cell and (b) Input phase response of the unit cells [23].

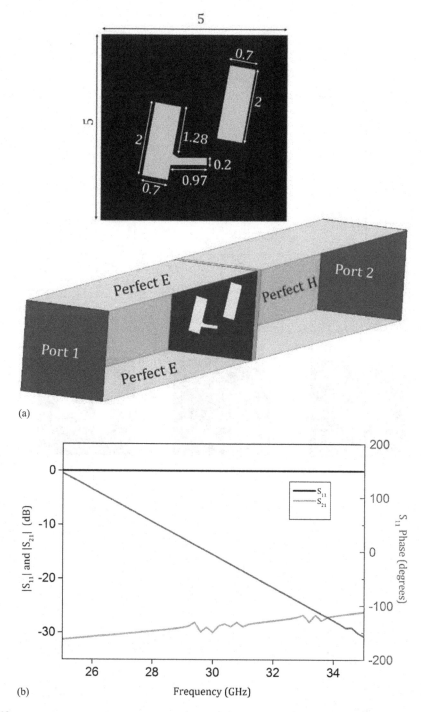

FIGURE 2.13
(a) Proposed unit cell with simulation model and (b) Simulated S11 and |S21| of the proposed unit cell [23].

overall size of the antenna. The overall size of the proposed unit cell is 0.5λ × 0.5λ at 28 GHz. Two slanted stubs are etched out of copper on the top plane of a Nelco NY9220 20 mil substrate.

The topology yields a linear phase ±90° in the frequency range of 27–33 GHz as depicted in Figure 2.13(b). Periodic boundary conditions were used in the simulations. The polarization of the incoming plane wave was identical to that of the antenna; also, the length of the waveguide was optimized to support only the dominant mode of propagation. The transmission and reflection characteristics are also illustrated in Figure 2.13(b). $|S_{21}|$ is below −25 dB in the 25–35 GHz band. A 2 × 4 array of unit cells was designed as the reflector. This translates to 1λ × 2λ at 28 GHz.

The reported designs of wideband reflectors have an aperture of more than 2λ × 2λ. The height of the proposed reflector decides the front-to-back ratio in the E-plane, a compromise between the pattern and the footprint of the antenna. The width of the reflector decides the H-plane pattern, and hence the forward gain. An offset of 1 mm is maintained between the CPW line and the reflector to prevent coupling. The proposed antenna backed by wideband reflector is shown in Figure 2.14(a).

The complete radiating structure is working in the quasi-waveguide mode, where the electrically small aperture radiates towards the reflector placed at 0.135λ. This undergoes multiple reflections followed by radiation from the effective aperture, as illustrated in Figure 2.14(b). The offset between the reflector and the radiating aperture is the critical parameter that decides the gain and impedance bandwidth. It is observed that when the reflector is electrically close to the radiator, the reflector acts as the primary radiator and the slot becomes a parasitic, hence the antenna is strongly detuned, since the reflector does not support radiation per se. When the reflector is electrically far away from the slot (offset being 3 mm), the impedance bandwidth increases to 28% with a deterioration in gain. The offset of 1.5 mm is chosen by considering the reflection coefficient and gain trade-off. A similar performance was observed when a perfect electric conductor (PEC) was substituted with a similar aperture, but with an offset at 2 mm, hence proving that the proposed reflector is relatively compact.

The simulated and measured reflection coefficient is depicted in Figure 2.15. The impedance bandwidth is from 27 to 33 GHz (20%). The slight discrepancy is caused by the non-ideal alignment between the reflector and radiator. The reflector acts as a shunt admittance to the radiator, hence offering an input impedance of $(39 - j4)\,\Omega$ and $(49 - j9)\,\Omega$ at 28 GHz and 31 GHz, respectively, but since the frequencies are close a wide impedance bandwidth is achieved.

Figure 2.16 illustrates the E-field plot of the side view, without and with reflector of the corner bent antenna. The transmission mode of the CPW line and the radiation mode is clearly visible in both the plots. It is also observed that without the reflector, the orthogonal ground plane splits the beam in both the directions. The diffraction effect with the reflector is also clearly visible.

The experimental setup used in the measurements is depicted in Figure 2.17. The inset photograph shows the zoomed view of the antenna under test. The simulated and measured co-pol and cross-pol patterns from 27 to 30 GHz are illustrated in Figure 2.18. The beamwidth in the E-plane (XZ plane) is around 76° ± 5°. The decrease in beamwidth is caused by the cavity effect of the reflector and the slot. The radiation

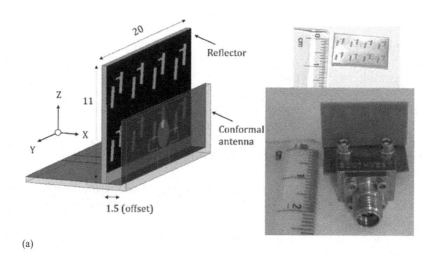

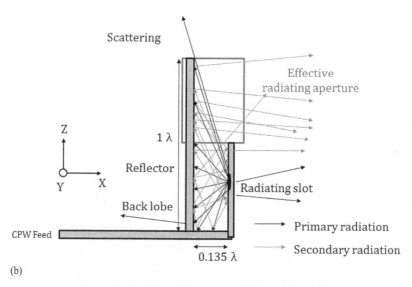

FIGURE 2.14
(a) Corner bent antenna with reflector design and (b) Field diagram of the radiation mechanism [23].

patterns are stable over the entire 20% band. The front-to-back ratio is more than 12 dB, as against 1 dB without the reflector. This ratio could be further improved with a larger effective aperture symmetrically, leading to an increase in the footprint of the antenna. These values prove the utility of the proposed antenna in the reduction in the specific absorption rate when integrated with a mobile terminal. The sidelobe at 190°–220° is primarily a result of the reduced aperture in the Z direction. The sidelobe suppression would increase the forward gain by 0.5 dB, hence the aperture is compromised. An L-bent reflector was also attempted to reduce the sidelobe, but it led to increased scattering from the sides. The patterns in H-plane (XY plane) have a beamwidth of 60° ± 5°. The front-to-back ratio is more than 14 dB for a 20% bandwidth. Since the effective aperture of the reflector is 2λ in the H-plane, a reduced beamwidth and superior

Conformal Antennas for Mobile Terminals 39

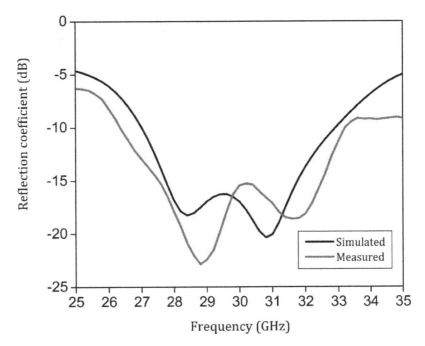

FIGURE 2.15
Reflection coefficient of corner bent antenna backed by wideband reflector [23].

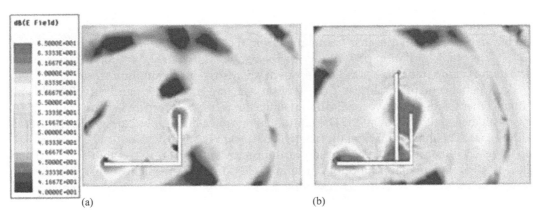

FIGURE 2.16
E-field plot of corner bent antenna (a) without and (b) with wideband reflector at 28 GHz [23].

front-to-back ratio is observed. The beamwidth increases at the higher end of the band because of the chosen offset length of 0.135λ.

The 3D patterns in Figure 2.19 provide an insight into the radiation without and with a reflector at 28 GHz; behaviour is similar across the entire band. The forward gain is shown in Figure 2.20 with the offset parametric analysis. As can be observed, the gain is highest when the reflector is offset at 1.5 mm. The parametric analysis is also in accordance with the gain-bandwidth principle. Since the increase in volume between the reflector and the radiating aperture leads to an increase in the impedance bandwidth and a consequent

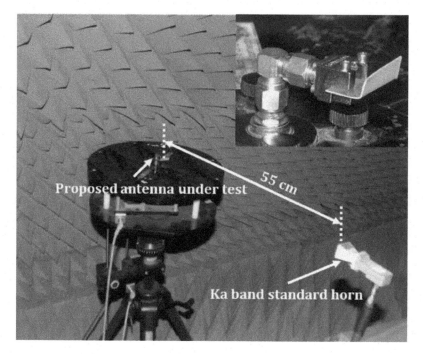

FIGURE 2.17
Experimental setup of radiation pattern measurement [23].

decrease in the gain. Simulated gain varies from 6 to 7 dBi in the frequency band from 25 to 35 GHz. Stable radiation patterns across the band resulted in stable gain in the H-plane. The gain is comparable to a strongly resonant microstrip patch antenna.

Gain enhancement of almost 2 dB is observed across the 20% band. Ideally, the gain enhancement should have been 3 dB, but since the reflector aperture is compromised, the gain enhancement has deviated. The simulated and measured gain curves are depicted in Figure 2.21. The maximum deviation between the two is 1.5 dB.

Table 2.1 illustrates the comparison of the presented design with other reported articles. A CPW-fed planar monopole antenna with dual beam is presented in [19]. The design has a couple of lumped resistors, which aid in gaining control of the antenna. Even though the design has a CPW feed, the primary radiator is over-moded and thus leading to a dual beam, which might not be a useful pattern for any use case scenario. Also, when actual lumped resistors would be integrated on to the antenna it would offer additional lumped capacitance and inductances at 28 GHz and beyond, and deteriorating the gain in addition to detuning the antenna. Also, designing an active tuning network that achieves a tuning from 500 Ω to 10 kΩ would be intricate and needs packaged integrated chips to do this.

A similar dual beam is also presented in [20], where planar metamaterial unit cells are integrated with the physical aperture of a stepped impedance transformer-fed bow-tie antenna. The gain enhancement observed is close to 2–3 dB, along with transformation from single beam to dual beam. The application of sub-wavelength unit cells is to change the effective refractive index of the medium of propagation of the waves emanating from the antenna. Additional metamaterial unit cells were used for reflecting action. This design strategy might not be applicable for the mobile phone usage scenario, as the antenna is end-fire and the gain enhancement principle used in end-fire would not work in the broadside. Also, this scheme lacks the concept of impedance bandwidth enhancement.

A high efficiency angled dipole is designed at 24 GHz in [21]. The principle of introducing an angle to the primary radiating arms of the dipole would increase the impedance bandwidth. This article also gives an exhaustive explanation on the choice of substrate and of characteristic impedance for the feed lines. Extension of the same design is reported in [22], where a coplanar stripline (CPS) feeding is used for feeding a folded dipole. Typically, the backend circuit feeding the antenna would be either microstrip based or CPW based, so additional baluns or an impedance transformer has to be integrated for any practical deployment.

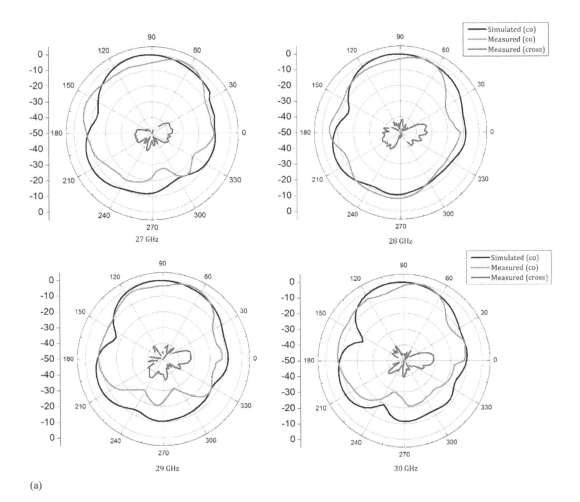

(a)

FIGURE 2.18
(a) Patterns in the XZ plane over 27–30 GHz frequency band
(Continued)

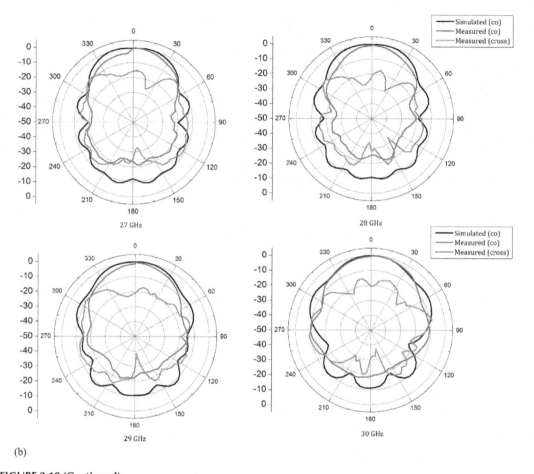

FIGURE 2.18 (Continued)
(b) Patterns in the XY plane over the 27–30 GHz frequency band [23].

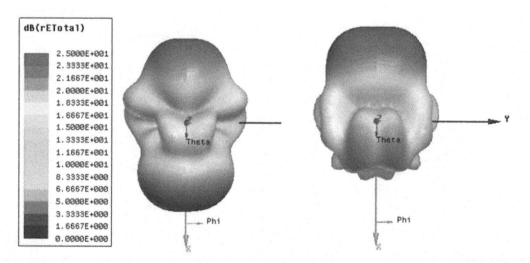

FIGURE 2.19
3D radiation patterns without and with reflector at 28 GHz [23].

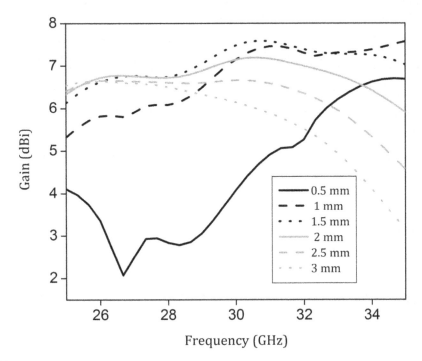

FIGURE 2.20
Simulated forward gain with parametric analysis [23].

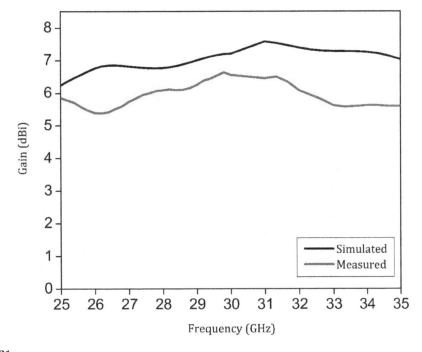

FIGURE 2.21
Forward gain of the corner bent antenna with reflector [23].

TABLE 2.1
Comparison with other Reported Articles

Ref.	Imp BW (GHz)	Gain (dBi)	Feed	Corner Bent
[19]	24–28 (15%)	3–4.5	CPW	No
[20]	57–64 (11.6%)	8–9	Microstrip	No
[21]	22–24 (8.7%)	4–6	Microstrip	No
[22]	21–25 (17%)	8–10	CPS	No
Proposed	27–33 (20%)	6–7	CPW	Yes

2.4 A Wideband High Gain Conformal Antenna for mmWave 5G Smartphones

The schematic of the proposed CPW-fed wideband mmWave antenna is illustrated in Figure 2.22. It is designed on a Nelco NY9220 substrate with a relative dielectric constant of 2.2, loss tangent of 0.0009 and thickness of 508 μm. The CPW feed line width and the gap were chosen to accommodate the end-launch connector and to prevent additional radiation resulting from an over-moded antenna element. Since the feed line with a width of 2.7 mm and a gap of 0.2 mm leads to a characteristic impedance of 59 Ω, a stepped impedance transformer of 50 Ω was utilized as an impedance transformer to high impedance of the radiating circular slot. Since the circular slot is uniformly illuminated independently of the

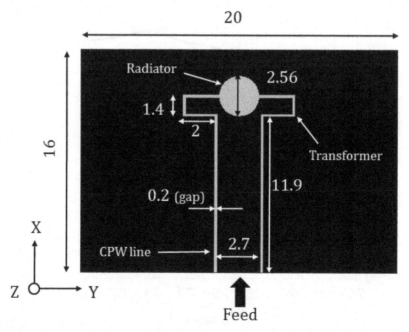

FIGURE 2.22
Proposed CPW-fed planar antenna (all dimensions are in mm) [23].

frequency, a wide impedance bandwidth is achieved. The impedance bandwidth of the planar element is from 20 to 28 GHz (30%)

The radiation is on both the sides of the element, since the antenna is not backed by an electrically large ground plane. Even though the antenna exhibits high pattern integrity, the front-to-back ratio is close to 0 dB throughout the band, consequently leading to a gain of 2–3 dBi, which might not be suitable for integration in mmWave 5G smartphones, which have requirements of high gain with minimal radiation towards the user. The designed CPW-fed antenna is planar, leading to a poor gain for the available radiating aperture of the smartphone, hence a 90° bend is introduced to make it conformal. The impedance bandwidth of the conformal antenna is from 28.5 to 33.5 GHz (16.2%), the shift in the operational band is caused primarily by the discontinuity introduced by the corner bending. The conformal antenna also radiates in both the directions as it is not backed by a conductor, but the front-to-back ratio has improved to 1 dB across the band as a result of the orthogonal ground plane. The forward gain of the conformal antenna is 4–5.6 dBi.

Thus, in order to improve gain and also to reduce the back radiation towards the user, an exponentially tapered reflector is integrated with the conformal radiator, as shown in Figure 2.23. The simplest way to achieve unidirectional radiation is to back the slot radiator with an electrically large copper flat plate at 0.25λ, but this method would yield a lower bandwidth and poor gain yield. Hence an exponentially tapered copper film was designed at a distance of 0.046λ from the slot radiator. A 3D-printed scaffolding was designed and fabricated using Raise 3D RxP2200 3D printer with a surface roughness of close to ±150 μm

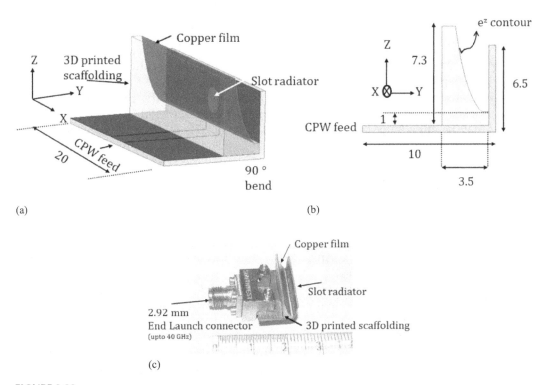

FIGURE 2.23
Proposed conformal antenna: (a) Schematic of the conformal antenna (isometric view), (b) Schematic of the conformal antenna (side view) and (c) Photograph of the fabricated prototype [23].

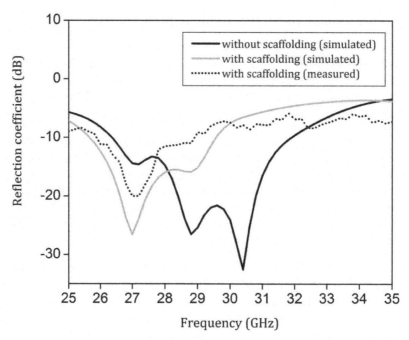

FIGURE 2.24
Input reflection coefficients of the conformal antenna [23].

and a polylactic acid (PLA) material with a dielectric constant of 2.75. The contour of the 3D-printed scaffolding was optimized for high gain, and wideband for an electrically close design. A thin copper film was deposited on to the 3D-printed scaffolding to realize the exponentially tapered reflector.

The simulated and measured input reflection coefficients of the proposed conformal antenna backed by an exponentially tapered reflector is depicted in Figure 2.24. It is observed that the impedance bandwidth is from 27 to 32 GHz (17%) when the slot radiator is backed by the exponentially tapered reflector alone. The impedance bandwidth detunes to 25–30 GHz (18.2%) primarily because of the coupling of the 3D-printed scaffolding with the CPW feed line. The bandwidth could be further enhanced by increasing the radiating volume between the slot radiator and the reflector, with a compromise in the physical footprint and forward gain.

The discrepancy between the simulated and measured curves is because of the non-ideal transition between the end-launch connector and the fabricated antenna element. All the simulations were performed in Ansys HFSS and the corresponding S-parameter measurements with Agilent PNA E8364C.

The 3D patterns with and without the reflector at 28 GHz are shown in Figure 2.25(a), it is evident that the reflector is effective in improving the front-to-back ratio, consequently leading to an enhancement of forward gain. The radiation patterns are illustrated in Figure 2.25, the beamwidth in the YZ plane is 65° ± 3° and in the XY plane 55° ± 5°, with a front-to-back ratio of more than 12 dB across the band. Since the beamwidth variation is minimal in both the principal planes, the pattern integrity is pretty high. The forward gain is shown in Figure 2.26.

The gain varies between 8–9 dBi across 25–35.5 GHz (34.7%) bandwidth, indicating a high gain yield for the effective radiating volume of $0.24\lambda^3$ at 28 GHz. The figure also

Conformal Antennas for Mobile Terminals

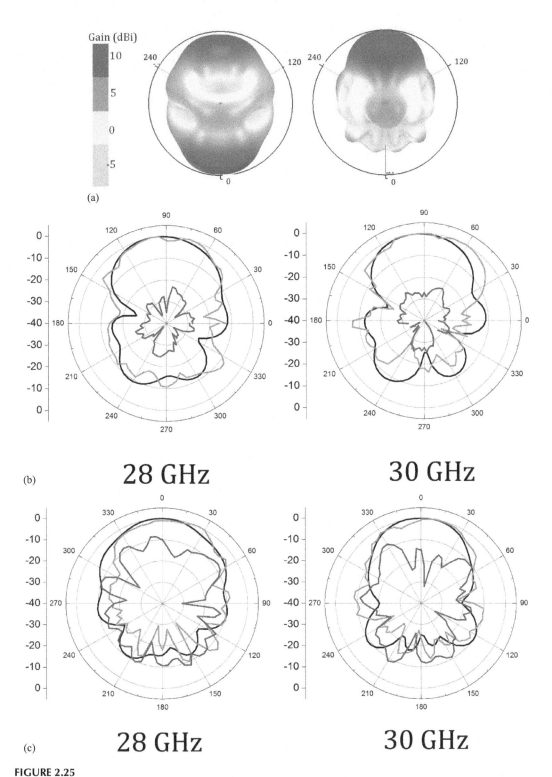

FIGURE 2.25
Radiation patterns of the conformal antenna, (a) 3D patterns without and with reflector at 28 GHz, (b) Patterns at 28 and 30 GHz in YZ plane (E-plane) and (c) Patterns at 28 and 30 GHz in XY plane (H-plane) [23].

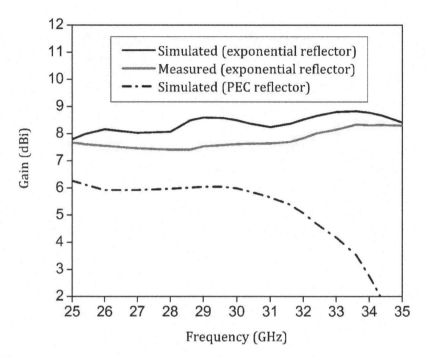

FIGURE 2.26
Forward gain of the conformal antenna [23].

demonstrates the forward gain when the slot radiator is backed by a conventional metal reflector at 0.25λ, which results in poor gain bandwidth. The gain could be further enhanced by increasing the height of the reflector, but this would compromise the physical footprint of the antenna.

Table 2.2 displays the advantages of the presented design compared to reported articles and reports. A wideband dipole array is presented in [3], where the reflecting ground plane is nearly 0.2λ from the radiator, hence achieving a 3 dB gain enhancement from the ground reflection with an improvement in the front-to-back ratio. The gain is compromised in spite of the half-wavelength dipole design. An extension of the element in the phased array

TABLE 2.2

Comparison with other Reported Articles

Ref.	F	IBW	G	GBW	ERV	Feed	Conformal
[3]	28	36.2	6	24.2	0.01	Microstrip	No
[4]	28	43	9	24.6	0.138	SIW	No
[5]	60	23.7	11	11.6	0.716	Microstrip	No
[6]	60	11.6	12	11.6	1.09	Microstrip	No
[7]	60	11.6	20	21.8	68.64	Microstrip	No
[8]	64	14.6	11	6.1	0.08	Microstrip	No
[9]	28	18.2	11	10.9	0.05	Microstrip	No
Proposed	28	18.2	9	34.7	0.24	CPW	Yes

Conformal Antennas for Mobile Terminals

topology is also presented. But an increase in the number of elements would eventually lead to an increase in effective aperture and subsequent increase in the gain of the phased array. A corner bent phased array is presented in Chapter 8 of this book. A substrate integrated waveguide (SIW)-fed metallic tapered slot antenna is illustrated in [4]. The radiation efficiency and the power handling capability of the antenna would be greater than typical dielectric based antennas, hence might not be relevant in the context of relatively low power antennas on the mobile terminal. Variants of metamaterial lenses are reported in [5,6], the concept of gain enhancement is to alter the effective refractive index of the physical aperture to achieve phase correction, and both the designs radiate in the end-fire. An electrically large dielectric lens is illustrated in [7] with significant gain. The occupied electrical size, when translated to 28 GHz, would be unsuitable for mobile terminals. It must also be noted that when the gain is beyond 20 dBi, the coverage would be very limited and might not be a clever solution for the mobile terminal.

A planar metamaterial unit cell integration is shown in [8] with low aperture efficiency, even though a gain enhancement of 2 dB is achieved across the band. A conventional Vivaldi antenna would require elongated apertures to achieve decent gain and impedance bandwidths. An interesting shared aperture design is demonstrated in [9], where the gain enhancement action is independent of the polarization of the incident wave. The limited beam scanning presented might be a solution for the base station but not for a smartphone ecosystem.

2.5 Design Guidelines for CPW-fed Conformal Antennas at Ka Band

The following generic design guidelines would aid in the design of conformal antennas at Ka band.

1. The choice of substrate for the planar antenna design is critical, since this influences the impedance and the radiation characteristics of the antenna at Ka band. The dielectric constant must be in the range of 2–3 to minimize surface wave modes. It must also be noted that a higher dielectric constant would lead to a relatively thinner trace width to realize the 50 Ω line, but the additional surface wave modes within the substrate would decrease the gain. The dielectric loss tangent must also be as low as possible to achieve maximum gain, if the antenna is tuned for the desired band. The thickness of the substrate must be in the range 10–20 mil, since thinner substrates would require additional scaffolding, decreasing the gain, whereas thicker substrates would lead to surface wave modes. It must also be noted that the choice of the substrate parameters must yield a feasible CPW line that could be fabricated, and the trace width must be less than 5 mm to avoid shorting of the trace pin and the grounding clamps of the end-launch connector.
2. Conformity of the antenna could be introduced. The bend must be introduced before the radiator, but the physical footprint of the radiator alone must be minimal to fit inside a typical mobile terminal. The antenna must be carefully bent to minimize the discontinuity of the conformal antenna. The introduction of the corner bend must be optimized in such a way as to create the least variation in the impedance characteristics.

3. Since the conformal antenna is not backed by a metallic ground plane, an additional reflector must be incorporated electrically close to the radiating aperture of the conformal antenna. The reflector must have a linear phase of $|S_{21}|$ (±90°) in the operating band. The $|S_{11}|$ must be close to 0 dB, indicating a high magnitude of reflection. The $|S_{21}|$ must be less than −10 dB, indicating a low leakage from the reflector. The designed reflector must be optimized after integration with the conformal antenna, since a clearance distance must be maintained from the feed line and the periodic geometry. A few design tweaks might be necessary post-integration of the reflector depending on the desired bandwidth, physical footprint and gain.

2.6 Conclusion

In this chapter, CPW-fed wideband conformal antennas were presented. The first design is a wideband corner bent antenna backed by an electrically close FSS reflector, and the second is backed by an exponentially tapered all-metallic reflector to yield high gain with high pattern integrity across the band. The proposed antennas could be a potential candidate for future 5G mobile terminal applications.

References

1. Forecast, Global Mobile Data Traffic. "Cisco visual networking index: global mobile data traffic forecast update 2017–2022," Update 2017 (2019): 2022.
2. C.-X. Wang et al. "Cellular architecture and key technologies for 5G wireless communication networks," *IEEE Commun. Mag.*, vol. 52, pp. 122–130, Feb. 2014.
3. S. X. Ta, H. Choo, and I. Park, "Broadband printed-dipole antenna and its arrays for 5G applications," *IEEE Antennas Wireless Propag. Lett.*, vol. 16, pp. 2183–2186, 2017.
4. B. Yang, Z. Yu, Y. Dong, J. Zhou, and W. Hong, "Compact tapered slot antenna array for 5G millimeter-wave massive MIMO systems," *IEEE Trans. Antennas Propag.*, vol. 65, no. 12, pp. 6721–6727, Dec. 2017.
5. A. Dadgarpour, B. Zarghooni, B. S. Virdee, and T. A. Denidni, "Improvement of gain and elevation tilt angle using metamaterial loading for millimeter-wave applications," *IEEE Antennas Wireless Propag. Lett.*, vol. 15, pp. 418–420, 2016.
6. A. Dadgarpour, B. Zarghooni, B. S. Virdee, and T. A. Denidni, "One- and two-dimensional beam-switching antenna for millimeter-wave MIMO spplications," *IEEE Trans. Antennas Propag.*, vol. 64, no. 2, pp. 564–573, Feb. 2016.
7. Z. Briqech, A. Sebak, and T. A. Denidni, "Wide-scan MSC-AFTSA array-fed grooved spherical lens antenna for millimeter-wave MIMO applications," *IEEE Trans. Antennas Propag.*, vol. 64, no. 7, pp. 2971–2980, July 2016.
8. M. Sun, Z. N. Chen, and X. Qing, "Gain enhancement of 60-GHz antipodal tapered slot antenna using zero-index metamaterial," *IEEE Trans. Antennas Propag.*, vol. 61, no. 4, pp. 1741–1746, Apr. 2013.
9. Z. Wani, M. P. Abegaonkar, and S. K. Koul, "Millimeter-wave antenna with wide-scan angle radiation characteristics for MIMO applications," *Int. J. RF Microw. Comput. Aided Eng.*, vol. 29, no. 5, e21564, 2019.

10. P. Choubey, W. Hong, Z.-C. Hao, P. Chen, T.-V. Duong, and M. Jiang, "A wideband dual-mode SIW cavity-backed triangular-complimentary-split-ring-slot (TCSRS) antenna," *IEEE Trans. Antennas Propag.*, vol. 64, no. 6, pp. 2541–2545, 2016.
11. K. Sarabandi, J. Oh, L. Pierce, K. Shivakumar, and S. Lingaiah, "Lightweight conformal antennas for robotic flapping flyers," *IEEE Antennas Propag. Mag.*, vol. 56, no. 6, pp. 29–40, Dec. 2014.
12. Y. J. Cheng, H. Xu, D. Ma, J. Wu, L. Wang, and Y. Fan, "Millimeter-wave shaped-beam substrate integrated conformal array antenna," *IEEE Trans. Antennas Propag.*, vol. 61, no. 9, pp. 4558–4566, Sept. 2013.
13. V. Semkin et al., "Beam switching conformal antenna array for mm-wave communications," *IEEE Antennas Wireless Propag. Lett.*, vol. 15, pp. 28–31, 2016.
14. L.-M. Si, W. Zhu, and H.-J. Sun, "A compact planar and CPW-fed metamaterial-inspired dual-band antenna," *IEEE Antennas Wireless Propag. Lett.*, vol. 12, pp. 305–308, 2013.
15. S. Raman and G. M. Rebeiz, "94 GHz slot-ring antennas for monopulse applications," *Antennas Propag. Soc. Int. Symp. Dig.*, vol. 1, pp. 722–725, June 1995.
16. G. Zhai, Y. Cheng, Q. Yin, S. Zhu, and J. Gao, "Uniplanar millimeter-wave log-periodic dipole array antenna fed by coplanar waveguide," *Int. J. Antennas Propag.*, 2013.
17. D. M. Elsheakh and M. F. Iskander, "Circularly polarized triband printed quasi-Yagi antenna for millimeter-wave applications," *Int. J. Antennas Propag.*, vol. 2015, pp. 1–9, Jan. 2015.
18. R. W. Jackson, "Considerations in the use of coplanar waveguide for millimeter-wave integrated circuits," *IEEE Trans. Microw. Theory Techn.*, vol. MTT-34, no. 12, pp. 1450–1456, Dec. 1986.
19. S. F. Jilani, S. M. Abbas, K. P. Esselle and A. Alomainy, "Millimeter-wave frequency reconfigurable T-shaped antenna for 5G networks," in *IEEE 11th Int. Conf. Wireless Mobile Comput. Netw. Commun. (WiMob)*, Abu Dhabi, 2015, pp. 100–102.
20. A. Dadgarpour B. Zarghooni B. S. Virdee, and T. A. Denidni, "Single end-fire antenna for dual-beam and broad beamwidth operation at 60 GHz by artificially modifying the permittivity of the antenna substrate," *IEEE Trans. Antennas Propag.*, vol. 64, no. 9, pp. 4068–4073, Sept. 2016.
21. R. A. Alhalabi and G. M. Rebeiz, "High-efficiency angled-dipole antennas for millimeter-wave phased array applications," *IEEE Trans. Antennas Propag.*, vol. 56, no. 10, pp. 3136–3142, Oct. 2008.
22. R. A. Alhalabi and G. M. Rebeiz, "Differentially-fed millimeter-wave Yagi-Uda antennas with folded dipole feed," *IEEE Trans. Antennas Propag.*, vol. 58, no. 3, pp. 966–969, Mar. 2010.
23. G. S. Karthikeya, M. P. Abegaonkar, and S. K. Koul, "CPW Fed Wideband Corner Bent Antenna for 5G Mobile Terminals," *IEEE Access*, vol. 7, pp. 10967–10975, 2019.

3

Flexible Antennas for Mobile Terminals

3.1 Introduction

In this chapter, flexible substrates for designing antennas at millimeter wave frequencies will be presented. Flexible antennas are an interesting class of antennas, where the characteristics of the antenna remain invariant even when its geometry is deformed. Flexible antennas have been investigated thoroughly in the lower frequency spectrum, especially for wearable applications and defence applications. Typically, in wearable applications, the antenna designed on an electrically and physically thin substrate would be conformed to the garment. The idea of designing a flexible antenna is to first choose a suitable substrate which has elastic properties or whose dielectric characteristics does not alter with the application of stress on the substrate. The antenna design is little more challenging than simply choosing a flexible substrate as there is the metallic portion in the antenna, whose properties also have to be investigated to understand the suitability of designing a flexible antenna. Flexible antennas might be one solution for antenna designs for mobile terminals. Designs with technical justification are presented in this chapter.

Typical data usage modes of a smartphone are presented initially to understand the design requirements. The primary use case of mmWave 5G is in the data usage mode, which needs antennas for the same application. Also, the typical usage scenarios presented in Chapter 2 must also be considered for conceiving the aforementioned designs. In accordance with these data usage modes, an orthogonal pattern diversity module is desired within which beams could be switched in orthogonal directions to cater to the modes. The antennas presented in the previous chapter have an additional reflector to achieve reasonable gain in the forward direction, which leads to an intricate design process and might lead to poor performance stemming from multiple components assembled together. The beam switching principle was not presented in the previous chapter. The radiating elements to be presented in this chapter would be for a specific application of data usage mode for mmWave. In this chapter, designs that are conformal but without reflectors will be presented along with a compact beam-switching module with high pattern integrity and low mutual coupling across the band. The validity of the design for the application scenario will also be clearly presented regarding the beam-forming corresponding to the modes of operation.

The typical usage modes of smartphones are portrait and landscape modes, as illustrated in Figure 3.1. The first mode is portrait mode, where the longer side of the mobile terminal is parallel to the axis of the head. This mode would generally be employed when a user wanted to check emails or surf the internet, for example. This mode is usually operated by one hand, with the smartphone being held in the other. The antenna placement in this situation is straightforward – it must be placed on the opposite edge of the smartphone so that radiation is minimal towards the user, but with the constraint that the mobile antenna must

FIGURE 3.1
Portrait and Landscape modes of operation [16].

offer high gain and comply with the dimensions of the panel. The second mode is landscape mode, and here the longer side of the smartphone is perpendicular to the axis of the user's head. As the centre of gravity is shifted, users tend to use both hands in this mode, making antenna placement for the landscape mode trickier than in the portrait mode, since both of the user's hands affect the signal. Popular design approaches such as phased arrays would suffer from severe scanning loss when excited in orthogonal modes, leading to unequal gains in both the modes. The concept of phased array and its associated failure in the context of orthogonal modes will be discussed in subsequent chapters. Hence an orthogonal pattern diversity module is presented in this chapter, which demonstrates the application of the proposed antenna designs for the aforementioned application scenario. The shared ground design with pattern diversity would lead to a larger physical footprint, since the port to port distance between the antennas for the specific modes would be greater than 20 mm, thus leading to a larger footprint compared to the overlapped design strategy.

A flexible Vivaldi antenna is also presented, indicating wide bandwidth and high gain. The bending analysis of the same is investigated.

3.2 Overview of Flexible Substrates for mmWave Applications

Several approaches have been reported in attempts to design flexible antennas. The most common approach in the lower frequencies is to use an electrically thin substrate such as Rogers 5880 with 5–10 mil and wrap around the desired geometry. The primary concern with this method of antenna design is that the additional scaffolding to be designed would be typically lossy dielectric, which would in turn lead to detuning the antenna in addition to contributing to specular patterns. It must also be observed that the parasitic dielectric supporting the radiating element might be a standard industry practice, as the real estate within the mobile terminal is severely limited and introducing scaffolding might be a roundabout design flaw.

There have been research articles published about the use of cloth-based flexible antennas but this strategy would be very lossy because of the poor conductivity of the copper used in the circuit and the underlying lossy dielectric. The substrate to be chosen must be sturdy and offer a frequency independent dielectric constant with a minimal dielectric loss tangent. Some of the substrates and their associated pros and cons are briefly discussed. Polyethylene terephthalate (PET) substrate was used in [1] but the dielectric loss tangent of 0.022 led to a poor gain yield, and PET is expensive compared to other conventional substrates. For a standard inset fed patch antenna operating at 28 GHz, the boresight gain is 8 dBi when the dielectric loss tangent of the substrate is 0.0009, which corresponds to Nelco NY9220, which was demonstrated in the previous chapter; similarly, the boresight gain decreases to 5.7 dBi when PET is used as a substrate, for the same dielectric constant, because of a higher loss tangent. This phenomenon is demonstrated in Figure 3.2. Naturally, a higher dielectric loss tangent indicates surface wave modes leading to poor gain and consequently lower radiation efficiency. The ideal substrate must have zero dielectric loss tangent, which could be achieved by using air as the substrate. If air is used as a substrate then the antenna must be suspended and the flexibility of the design might be compromised. The air substrate or all-metallic designs are common in the lower frequency, so frequency scaling of these designs might be complicated and may be unrealizable. A few researchers have also reported using 3D-printed substrate with micromachining to improve radiation efficiency, but the flexibility of the design is questionable.

Thus, flexible, low dielectric constant, low dielectric loss, low-cost substrate is a preferred candidate for antenna designs at Ka band. It must be noted that the liquid crystal polymer (LCP) antenna illustrated in [2] requires an expensive substrate in addition to a complicated non-industry-standard manufacturing process indicating a poor cost-to-volume ratio. Even though the fabrication of the 3D-printed antenna demonstrated in [3] is low-cost with a low dielectric loss tangent, the selective metallization of the radiator requires time-consuming and expensive inkjet printing with silver thick-film paste. There are several other problems associated with conductive inkjet printing, such as periodic clogging and poor resolution. The ink used in this process must be customized for the application in hand and requires a massive chemistry drama to conceive the ink. Hence, antenna designs based on a flexible, low-cost polycarbonate substrate are proposed in this chapter. Industry-standard chemical etching is used for circuit fabrication on the substrate.

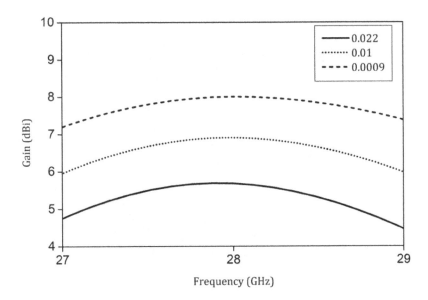

FIGURE 3.2
Broadside gain of a patch antenna with substrates with different dielectric loss tangent [16].

3.3 Corner Bent Patch Antenna for Portrait Mode

The schematic of the proposed corner bent inset fed patch antenna, along with the photograph of the fabricated prototype, is depicted in Figure 3.3; all the dimensions are in millimetres. The chosen substrate is 500 µm polycarbonate with ε_r of 2.9 and dielectric loss tangent of 0.01. The chosen substrate is a compromise between cost and flexibility; 250 µm thick substrates are also on the market, which would yield a higher gain because of a

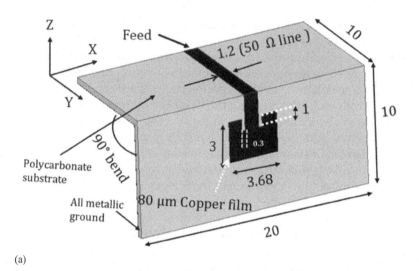

(a)

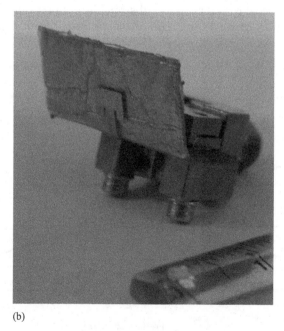

(b)

FIGURE 3.3
(a) Schematic of the corner bent patch antenna (units: mm) and (b) Photograph of the corner bent patch antenna [16].

lower energy loss in the substrate. However, the structural stability of thinner substrates is questionable. Polycarbonate is a universally available plastic and costs only US$8 for a sheet of 3 m × 1 m. Also, electrically thinner substrates with low dielectric constant require supporting structures, which would in turn compromise the physical footprint and the forward gain, because of an undesired coupling between the radiator and the intended parasitic all-dielectric scaffolding. The feed line of the proposed antenna is a standard 50 Ω line feeding the strongly resonant patch antenna. A 50 Ω line is feasible in the context of microstrip feeding technique, as the width of the line is only 1.2 mm, translating to 0.11λ at 28 GHz, and the possibility of radiation is minimized from the feed line. Also, the E-fields do not add in phase during the transmission line mode, thus reducing undesired radiation from the feeding structure. Ideally, the dielectric constant in the feeding section of the antenna must be high to confine the energy flow to the substrate, and the dielectric constant under the radiating section must be low to promote minimal energy loss in the substrate underneath. But realizing this would be intricate and hence is usually avoided. The width of the overall antenna was chosen to mount the Southwest end-launch connector, which is rated up to 40 GHz. Even though low-cost SMA connectors could be used in the context, the impedance mismatch between the trace of the SMA and that of the antenna would be higher than end-launch connectors. Also, polycarbonate based designs do not support soldering, as the melting point of the plastic dielectric is 155°C whereas the typical soldering temperature is around 180°C–200°C, indicating that when a copper trace over a polycarbonate substrate is soldered, the dielectric immediately beneath the feed line would melt more quickly than the solder, thus making it impossible to use soldering with connectors on polycarbonate. In the actual deployment scenario the feed lines originating from the circuitry could be wire-bonded to the antennas with suitable matching network introduced between the wire bond and the antenna trace. The chemical etching followed for conventional copper-clad substrates has to be tweaked for fabrication on polycarbonate. The reader can compare the fabrication process on conventional substrates in Chapter 10.

The steps for the fabrication of passive microwave circuits on a polycarbonate substrate are given below:

1. The required dimensions of the polycarbonate are cut – polycarbonate is supplied as large sheets and has to be cut to the dimensions required by the antenna. Precision laser cutting could be used for cutting the substrate. It is recommended to allow additional substrate compared to the size of the antenna in case of errors during fabrication. The resolution of the cutting machine is not a very critical parameter during this process.
2. Copper tape of thickness 80 μm is pasted carefully without air bubbles on both the sides of the substrate. Ideally, thinner copper films of less than 50 μm would also be operational considering the skin depth at this frequency of operation. Thicker films would lead to additional conductor losses, which might create a discontinuity with the substrate. As a demonstration to the proof of the concept, industry grade copper tape was used in the upcoming design.
3. Industry standard chemical etching process is carried out on the top plane. The fabrication process for conventional substrates has to be optimized to accommodate the copper-pasted polycarbonate. For example, the application of photoresist has to be done carefully on the copper film. Despite the copper pasted polycarbonate having structural stability, the copper film's behaviour needs to be investigated. It is

recommended to fabricate multiple samples to minimize errors in fabrication process. The critical aspect of fabrication is the amount of UV radiation to which the sample is exposed.

4. The edge of the feed line and the ground planes are filed for better contact with the high frequency connector. The copper film or paste might not be connector ready and might have some surface roughness which has to be reduced by manual filing or milling the edge of the connector. Acetone could also be used as a cleansing agent to remove additional dirt from the fabricated sample.

Since both the substrate and the process are inexpensive, the antenna would yield low cost per unit when mass produced. The radiator is at least 1λ away from the feed plane to reduce the effect of the electrically large connector [4–6]. The corner bend is introduced with the radiator, hence the transmission line lies in the orthogonal plane with respect to the radiator, thereby decreasing the physical footprint of the antenna post integration with the mobile panel. Since the substrate is flexible, minimal discontinuity was created in the overall geometry of the antenna, and copper tape was more flexible compared to the conventional chemical vapour deposited (CVD) substrate.

Since, the radiator is bent with the ground plane, there is no requirement for additional FSS based reflectors, as reported in sections 2.3 and 2.4. Thus this proved to be a simpler design and more robust than the previously illustrated design examples.

The input reflection coefficient is depicted in Figure 3.4. The impedance bandwidth is from 27 to 29 GHz, i.e., 7.2%. The variation between simulated and measured curves could be attributed to the discontinuity of the trace between the connector and the antenna. The dielectric constant of the polycarbonate substrate might have deviated from the values

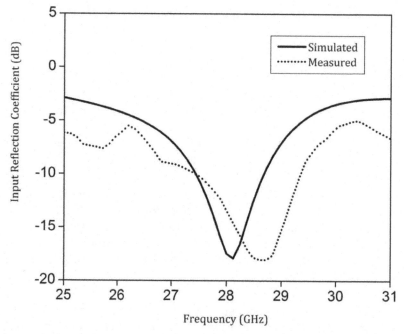

FIGURE 3.4
Reflection coefficient of corner bent patch antenna [16].

Flexible Antennas for Mobile Terminals

assumed in simulations. The bandwidth could be further increased by increasing the effective radiating volume, which would lead to a compromise in the physical footprint. Hence, the proposed design is a compact antenna with high forward gain, which is essential for antennas to be integrated with future mobile terminals. The S-parameter measurements were done using Agilent PNA E8364C. All the full-wave simulations were carried out using Ansys HFSS. The simulated and measured radiation patterns in the principal orthogonal plane cuts are shown in Figure 3.5. The beamwidth in the YZ plane is 65° and XY plane is 62° at 28 GHz, the front-to-back ratio is close to 18 dB, indicating the low value of radiating power towards the user when integrated with a mobile terminal.

The forward gain is close to 7 dBi at 28 GHz, as observed in Figure 3.6. The pattern integrity pre- and post-bending is almost identical, proving the utility of the flexible polycarbonate substrate.

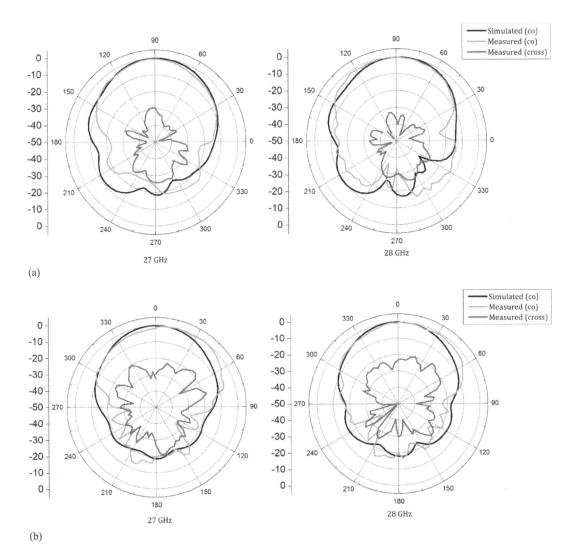

FIGURE 3.5
Radiations patterns at 27 and 28 GHz in (a) YZ and (b) XY planes [16].

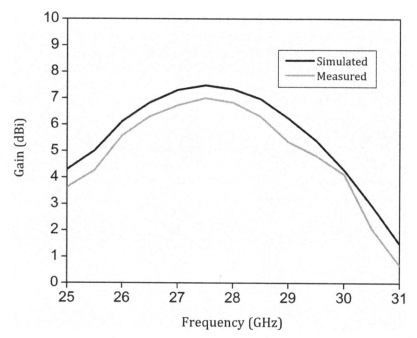

FIGURE 3.6
Forward gain of the proposed antenna [16].

3.4 Corner Bent Tapered Slot Antenna for Landscape Mode

The corner bent antenna presented in Section 3.3 would be operational for portrait mode. In order to cater to the landscape mode, a corner bent end fire radiator is investigated in this section. The proposed corner bent tapered slot antenna (TSA) along with a photograph of the fabricated prototype is illustrated in Figure 3.7. The feed is a standard 50 Ω line leading to the microstrip to slotline transition through a couple of stepped impedance transformers. The aperture is designed in such a way that the gain is reasonably high with least physical footprint. It must also be noted that since the aperture is optically transparent, solar operated module could also be integrated when the user operates the smartphone in landscape mode.

The simulated and measured input reflection coefficients are shown in Figure 3.8. The 10 dB impedance bandwidth is 25–29.5 GHz, translating to 16.5%. The bandwidth could be further increased by additional balun, but at the cost of deterioration in pattern integrity across the band. The discrepancy between simulated and measured curves might be related to manufacturing errors and the solder-free transition from the end-launch connector to the fabricated element.

The radiation patterns are shown in Figure 3.9 in both the principal cuts of the corner bent TSA. The front-to-back ratio is greater than 10 dB because of the electrically large ground plane, thus preventing radiation towards the user when integrated with a mobile terminal.

The forward gain is illustrated in Figure 3.10, indicating a gain of 7 dBi at 28 GHz. The variation between simulated and measured gains might be a result of polarization alignment errors during gain measurement by the two-antenna method.

Flexible Antennas for Mobile Terminals

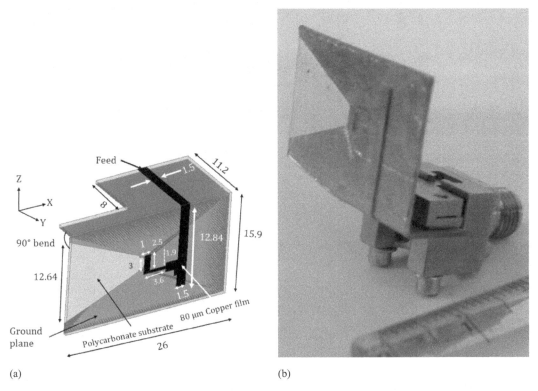

FIGURE 3.7
(a) Schematic of the corner bent tapered slot antenna (units: mm), and (b) Photograph of the corner bent tapered slot antenna [16].

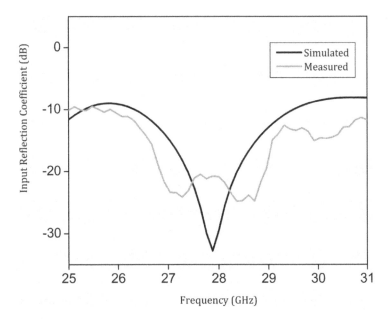

FIGURE 3.8
Reflection coefficient of corner bent TSA [16].

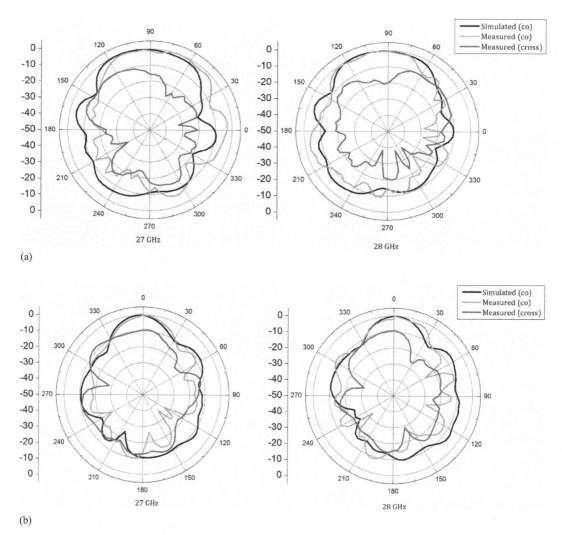

FIGURE 3.9
Radiations patterns at 27 and 28 GHz in (a) XY and (b) XZ planes [16].

To integrate both the proposed antennas, an overlapped architecture is demonstrated, as shown in Figure 3.11. Both the corner bent antennas are bonded via a 3D-printed scaffolding of 1.4 mm thickness, translating to a port-to-port distance of 0.13λ at 28 GHz with minimal distortion in the patterns when the corresponding port is excited. The mutual coupling between the electrically close ports is less than 27 dB, as can be seen in Figure 3.12.

The patterns retain their integrity in spite of electrically close integration, as is evident from the patterns at 28 GHz in Figure 3.13. Port 1 corresponds to the corner bent patch, and Port 2 to the tapered slot antenna. It must also be observed that since the ground plane of the patch antenna is operating as a natural decoupling network, the pattern distortion is minimal. Also, the E-plane pattern of the TSA is almost identical to the stand-alone element. However, the H-plane pattern is slightly deteriorated because of the obstructing radiator. Since the module is electrically compact, the compromise in patterns is justified.

Flexible Antennas for Mobile Terminals

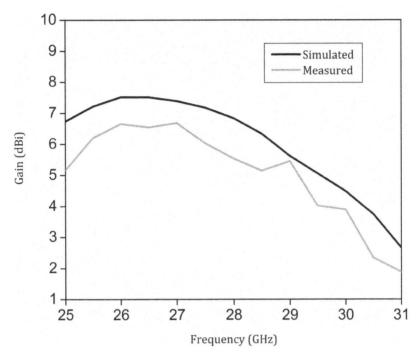

FIGURE 3.10
Forward gain of the proposed corner bent TSAs. Overlapped Orthogonal Pattern Diversity Module

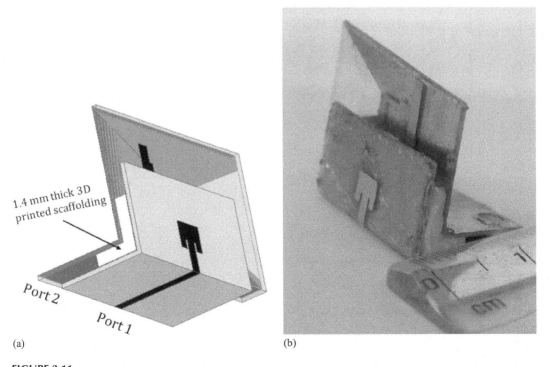

FIGURE 3.11
(a) Schematic for overlapped pattern diversity module, and (b) Photograph of the fabricated prototype [16].

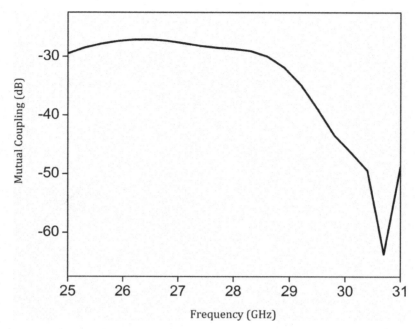

FIGURE 3.12
Mutual coupling of the overlapped module [16].

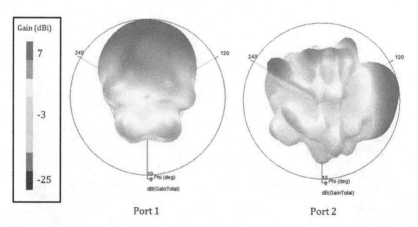

FIGURE 3.13
3D patterns of the overlapped module at 28 GHz [16].

Table 3.1 illustrates the design features of the presented design in comparison with reported designs. It is evident that the module occupies minimum volume with reasonably high gain with orthogonal pattern diversity. A SIW-fed metallic tapered slot antenna is reported in [7], even though a beam scanning in the range of ±40° is reported. The actual implementation of a wideband phase shifter integrated design would be intricate to implement, along with the necessary controllers. The scanning loss is close to 2 dB when a beam

TABLE 3.1

Comparison with Reported Articles

Ref.	F	G	Con.	MC	ERV	PD
[7]	28	9	No	NA	0.138	+40°, −40°
[8]	16	8.5	No	25	0.14	0°, 180°, ±90°
[9]	10	14	No	NA	0.04	No
[10]	2.5	6.1	No	42	0.116	No
[11]	5.8	4	No	20	0.016	0°, ±90°
[12]	60	12	No	20	1.09	+35°, −35°
[13]	60	20	No	30	68.64	0°, +30°, −30°
[14]	64	11	No	NA	0.08	No
[15]	28	11	No	16	0.05	0°, +30°, −30°
Proposed	28	7	Yes	30	0.12	0°, 90°

Note F = centre frequency (GHz), G = gain (dBi), Con. = conformal, MC = mutual coupling (dB), ERV = effective radiating volume (λ^3), PD = pattern diversity.

is scanned away from the 0° axis of its radiation. The weight of an all-metallic based design might also be incompatible with typical smartphones.

The physical footprint of the antenna and its array counterpart is pretty high. Orthogonal beams have been reported in [8] in the lower frequency, but the antenna module would have increasing blind spots at the higher end of the spectrum.

Also, the concept of cross-configuration would lead to increased electrical size, which in turn would lead to poor conformity with the traditional mobile environment.

Multi-layer metamaterial unit cells are integrated with a high gain wideband Vivaldi antenna in [9]; the multi-layer primarily aiding phase correction in both of the principal planes of the antenna in the end-fire orientation.

A similar concept could have been used in the tapered slot antenna of the presented antenna in Section 3.4, but the magnitude of gain enhancement at the cost of increased physical aperture is questionable. Also, the fabrication of sub-wavelength unit cells on a polycarbonate substrate might be complicated.

A metasurface based mutual coupling reduction strategy is presented in [10]. The additional design and integration of a metasurface would increase the footprint in the orthogonal direction for the presented overlapped module. The antennas presented in [10] have similar radiation characteristics compared to the orthogonal beam forming in the design presented.

An electrically compact multi-orthogonal pattern diversity module is presented in [11]. The mutual coupling reduction stems from the orthogonal beams, hence no additional circuitry for mutual coupling reduction is explored.

The topology of the antennas is designed in such a way that the electrical spacing does not significantly deteriorate the patterns of the corresponding ports. Also, the design is illustrated in a lower frequency, where a coaxial feeding technique is feasible, but the same might not be feasible at 28 GHz because of the electrical size of the connector feeding the radiating structure.

Multi-polarization antennas with variants of planar metamaterial unit cells are presented in [12–15]; similar topology might lead to poor gains when incident with orthogonal beams, thus proving to be incompatible with the application at hand.

3.5 Dielectric Loaded Polycarbonate Based Vivaldi Antenna

In this section, a planar dielectric loaded antenna is presented. In the first step, we designed a conventional antipodal Vivaldi antenna with a copper layer of thickness 0.017 mm at both sides, printed on a substrate of polycarbonate. The antenna was designed with the thickness of the substrate 0.5 mm, the dielectric constant (ε_r = 2.9), and loss of tangent (tan δ = 0.01). The overall dimensions of the antenna are approximately 35 × 20 × 0.5 mm³ (length × width × height). The feed-line width is 1.2 mm in order to match to the 50 Ω impedance. In a conventional AVA design structure, the aperture width should be equal to half of the wavelength at the desired frequency of operation. It is well-known that an exponential curve provides a better performance the antenna when compared to parabola or trigonometric functions. The antenna was designed by mainly using a simple coordinate system as shown below. Two endpoints, namely (x1, y1), (x2, y2) and r were set as variables. The exponential curves employed in the design can be described by

$$x = \left(ae^{ry} + b\right) \tag{3.1}$$

$$a = \frac{x1 - x2}{e^{ry1} - e^{ry2}} \tag{3.2}$$

$$b = x1 - ae^{ry1} \tag{3.3}$$

where x1, x2 are the abscissa, y1, y2 are the ordinates of two endpoints, r is the gradual change ratio, a, b are two parameters of the exponential function. The designed antenna shown in Figure 3.14 operates in the 24–35 GHz band centred at 28 GHz. The exponential profile curve is drawn using '1' where a, b parameters are found using '2', '3', (x1, y1) = (17.5, 9.4), (x2, y2) = (35, 10.6) which gives radius of curvatures as r1 = 0.09, r2 = 0.25.

Through simulation optimization, the optimal impedance matching is observed when the breadth of feed is 1.2 mm. However, the characteristic impedance is 52.3 Ω when the depth of the substrate is 0.5 mm and the breadth of feed is 1.2 mm. So only a certain impedance mismatch occurs when connected directly to the 50 Ω feed line. Here, the aperture width is 20 mm which is greater than λ_0, and length of the antenna is 6 × λ_0 where λ_0 is free space wavelength. Several common optimization methods are available. In this design, corrugations are used, of unequal length, which improves the return loss and increases the gain with good radiation patterns.

In the next step we loaded the antenna with a dielectric substrate in order to improve the properties of the antenna, which is largely used for gain enhancement. The antenna has overall dimensions of approximately 60 × 20 × 0.5 mm³ (length × width × height), i.e., shown in Figure 3.14(b) and (c). For the enhancement of gain, one of the most important factors is the shape of the load. Through simulation, a comparison between different shapes such as triangle, trapezium and ellipse were made, the proposed antenna with the basic antenna without the load was observed and the enhancement was increased because of the designed antenna, i.e., approximately 1.5 dB from 8.19 to 9.69 dB. In this chapter, the method used for the fabrication was chemical etching.

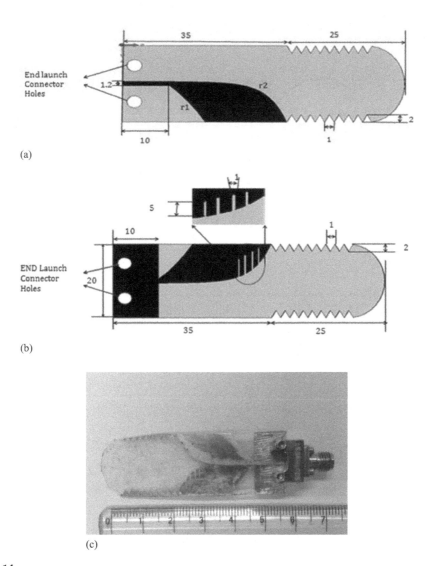

FIGURE 3.14
(a) Front of designed antenna (units: mm), (b) Schematic of the ground plane and (c) Photograph of the fabricated prototype.

Figure 3.15 shows simulation and measured results of the variation of reflection coefficient (S_{11}) with frequency for the designed antenna. The reflection coefficient of the conventional antipodal Vivaldi antenna is below −10 dB for the frequency range of 28–35 GHz. Application of the corrugation for the proposed antenna has improved the return loss, mainly at the centre frequency; after loading the S_{11} is less than −15 dB from 24 to 32 GHz (20% impedance bandwidth) and at the centre frequency it is −43 dB. The variation of realized gain against frequency is shown in Figure 3.16, with dielectric loaded antenna, and proposed antenna with no load. As can be observed from the figure, the loading of AVA with the corrugations improved gain performance throughout the operating frequency band of the proposed antenna.

Similarly, the increase in the effective length of the proposed antenna resulting from the modification leads to a more directive beam in both the E-plane and H-plane. The realized

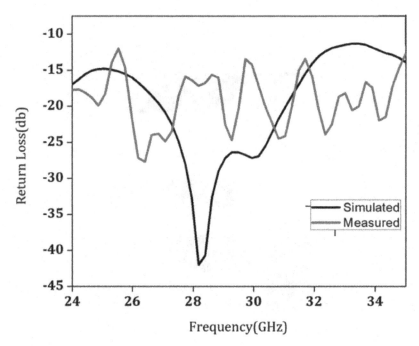

FIGURE 3.15
Simulated and measured input reflection coefficients.

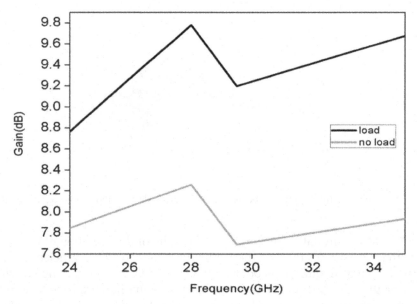

FIGURE 3.16
End-fire gain of the proposed antenna.

gain of the proposed AVA is found to be 7.6–8.3 dB over the 24–35 GHz bandwidth, while the realized gain of the proposed antenna with dielectric loading is between 8.7–9.8 dB.

Increase in gain can be observed throughout the operating band of the antenna, with the increase of 1.5 dB at the centre frequency 28 GHz. In Figure 3.17, different shapes of load are plotted, for example, ellipse, trapezium, triangle, and proposed antenna with the

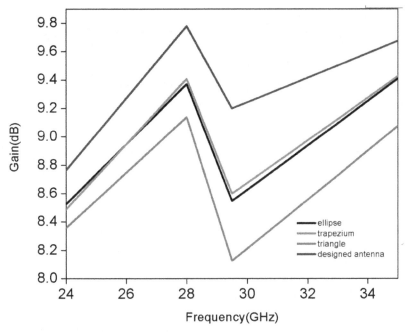

FIGURE 3.17
End-fire gains for various profiles of dielectric loading.

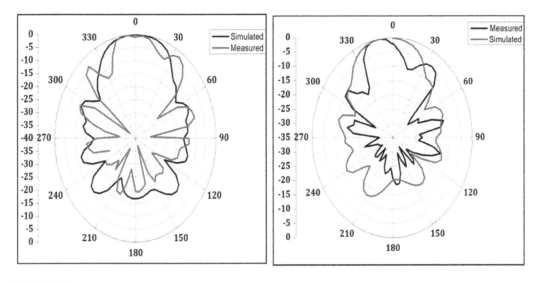

FIGURE 3.18
E-plane radiation patterns at 28 and 32 GHz.

variation of realized gain versus frequency is shown. It is observed that gain enhancement related to ellipse, trapezium, triangle and proposed antenna are 1.19, 1.21, 0.9, 1.5 dB, respectively. The E-plane radiation patterns for the frequencies of 28 and 32 GHz are shown in Figure 3.18, where there is a reasonable amount of dissimilarity between the simulated and measured radiation characteristics, primarily because of the effect of lossy adapters utilized for measurements.

FIGURE 3.19
E-field patterns without and with dielectric loading at 28 GHz.

E-field distribution related to loading: to study E-field distribution at 28 GHz because of the effect of loading on the antenna, we look at Figure 3.19, which indicates that as a result of loading the wavefront emerging out of the antenna is more planar than its unloaded counterpart.

The antenna is bent using a cylinder for three different values of radii, i.e., 10, 11 and 12 cm and is compared with the designed antenna. Return loss is evaluated and compared to check the shift/decrease caused by impedance mismatch and change in the effective electrical length of the radiating elements. When bent, the distortions that are also possible as a result of this gain are also tested. The return loss variation against various bending radii is observed in Figure 3.20, and its corresponding gains in Figure 3.21.

Hence a flexible high gain wideband Vivaldi antenna is presented which could be utilized for mmWave 5G cellular devices.

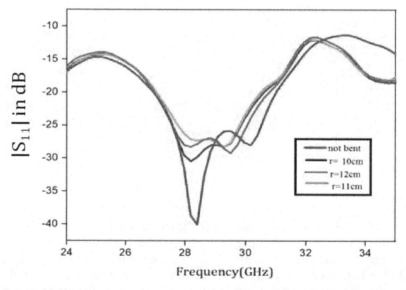

FIGURE 3.20
Input reflection coefficient variation with bending radii.

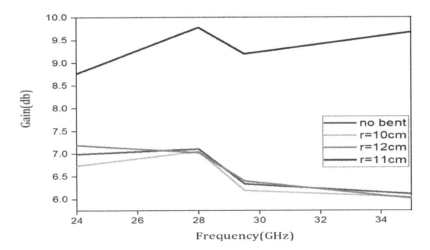

FIGURE 3.21
End-fire gain variations with bending radii.

3.6 Conclusion

In this chapter, corner bent inset fed patch antenna and tapered slot antenna operating at 28 GHz designed on a low-cost polycarbonate substrate was presented for future 5G mobile terminals. An industry standard low-cost chemical etching process was used for fabrication of the prototypes. An overlapped orthogonal pattern diversity module was also presented, which yields a gain of 7 dBi for both portrait and landscape modes of operation with a mutual coupling of less than 28 dB in spite of being electrically close to each other. Hence the proposed module might be a candidate for mass-produced 5G antennas. A flexible Vivaldi antenna was also presented, indicating wide bandwidth and high gain. The bending analysis of this was investigated.

References

1. S. F. Jilani and A. Alomainy, "Planar Millimeter-Wave Antenna on Low-Cost Flexible PET Substrate for 5G Applications," *2016 10th European Conference on Antennas and Propagation (EuCAP)*, Davos, 2016, pp. 1–3
2. S. F. Jilani, M. O. Munoz, Q. H. Abbasi, and A. Alomainy, "Millimeter-Wave Liquid Crystal Polymer Based Conformal Antenna Array for 5G Applications," *IEEE Antennas and Wireless Propagation Letters*, vol. 18, no. 1, pp. 84–88, Jan. 2019.
3. D. F. Hawatmeh, S. LeBlanc, P. I. Deffenbaugh, and T. Weller, "Embedded 6-GHz 3-D Printed Half-Wave Dipole Antenna," *IEEE Antennas and Wireless Propagation Letters*, vol. 16, pp. 145–148, 2017.
4. S. X. Ta, H. Choo, and I. Park, "Broadband Printed-Dipole Antenna and Its Arrays for 5G Applications," *IEEE Antennas and Wireless Propagation Letters*, vol. 16, pp. 2183–2186, 2017.

5. S. Zhu, H. Liu, Z. Chen, and P. Wen, "A Compact Gain-Enhanced Vivaldi Antenna Array with Suppressed Mutual Coupling for 5G mmWave Application," *IEEE Antennas and Wireless Propagation Letters*, vol. 17, no. 5, pp. 776–779, May 2018.
6. R. A. Alhalabi and G. M. Rebeiz, "High-Efficiency Angled-Dipole Antennas for Millimeter-Wave Phased Array Applications," *IEEE Transactions on Antennas and Propagation*, vol. 56, no. 10, pp. 3136–3142, 2008.
7. B. Yang, Z. Yu, Y. Dong, J. Zhou, and W. Hong, "Compact Tapered Slot Antenna Array for 5G Millimeter-Wave Massive MIMO Systems," *IEEE Transactions on Antennas and Propagation*, vol. 65, no. 12, pp. 6721–6727, Dec. 2017.
8. G. S. Reddy, A. Kamma, S. Kharche, J. Mukherjee, and S. K. Mishra, "Cross-Configured Directional UWB Antennas for Multidirectional Pattern Diversity Characteristics," *IEEE Transactions on Antennas and Propagation*, vol. 63, no. 2, pp. 853–858, Feb. 2015.
9. B. Zhou, H. Li, X. Zou, and T.-J. Cui, "Broadband and High-Gain Planar Vivaldi Antennas Based on Inhomogeneous Anisotropic Zero-Index Metamaterials," *Progress in Electromagnetics Research*, vol. 120, pp. 235–247, 2011.
10. F. Liu, J. Guo, L. Zhao, X. Shen, and Y. Yin, "A Meta-Surface Decoupling Method for Two Linear Polarized Antenna Array in Sub-6 GHz Base Station Applications," *IEEE Access*, vol. 7, pp. 2759–2768, 2019.
11. Y. Sharma, D. Sarkar, K. Saurav, and K. V. Srivastava, "Three-Element MIMO Antenna System With Pattern and Polarization Diversity for WLAN Applications," in *IEEE Antennas and Wireless Propagation Letters*, vol. 16, pp. 1163–1166, 2017
12. A. Dadgarpour, B. Zarghooni, B. S. Virdee, and T. A. Denidni, "One- and Two-Dimensional Beam-Switching Antenna for Millimeter-Wave MIMO Applications," *IEEE Transactions on Antennas and Propagation*, vol. 64, no. 2, pp. 564–573, Feb. 2016.
13. Z. Briqech, A. Sebak, and T. A. Denidni, "Wide-Scan MSC-AFTSA Array-Fed Grooved Spherical Lens Antenna for Millimeter-Wave MIMO Applications," *IEEE Transactions on Antennas and Propagation*, vol. 64, no. 7, pp. 2971–2980, July 2016.
14. M. Sun, Z. N. Chen, and X. Qing, "Gain Enhancement of 60-GHz Antipodal Tapered Slot Antenna Using Zero-Index Metamaterial," *IEEE Transactions on Antennas and Propagation*, vol. 61, no. 4, pp. 1741–1746, Apr. 2013.
15. Z. Wani, M. P. Abegaonkar, and S. K. Koul, "Millimeter-Wave Antenna with Wide-Scan Angle Radiation Characteristics for MIMO Applications," *Int. J. RF Microw. Comput. Aided Eng.*, vol. 29, no. 5, p. e21564, 2018.
16. G. S. Karthikeya, M. P. Abegaonkar, and S. K. Koul, "Polycarbonate Based Overlapped Architecture for Landscape and Portrait Modes of mmWave 5G Smartphone," *Progress in Electromagnetics Research M*, Vol. 86, 135–144, 2019.

4

Compact Antennas with Pattern Diversity

4.1 Introduction

In this chapter orthogonal pattern diversity antenna elements to support the landscape and portrait modes in a typical smartphone environment are presented. As discussed in previous chapters, millimeter wave 5G is targeted primarily at data usage, which means that the antennas to be designed must be able to support this. Statistically speaking, users would utilize the smartphone either in landscape or portrait mode. The antennas to be designed must radiate minimally towards the user, and the placement of the antenna module has to be in such a way that the mutual coupling among the elements must be as low as possible, and the elements have to be fired according to the mode of operation.

It is assumed that the switches which would be controlling the antennas would be activated when maximum power is received by one of the antennas in the integrated module. In the previous chapters, a corner bent broadside antenna was investigated, operating in the 28 GHz band. It was clear that a wideband high gain antenna initially radiating in the broadside orientation could be modified to radiate minimally towards the user and towards the base station. In this chapter, a planar antenna radiating in the end-fire orientation would be corner bent with a reflector to validate the strategy presented in the earlier chapters. The first design on wideband conformal antenna fed by CPW with an end-fire radiation pattern is presented. The end-fire antenna is bent immediately after the transformer and before the radiator. The patterns for the same are thoroughly studied and an electrically close reflector is integrated for gain enhancement and consequent decrease in SAR post integration. Earlier, it was mentioned that the reflector integration would lead to a reduction in the effective SAR; the details of this integration and its demonstration is presented in this chapter. The second design is a gain compensated pattern diversity module for mobile terminals. In this design, three antennas are presented initially with microstrip feeding technique and built on a substrate that has reasonable operational characteristics in the Ka band.

The microstrip fed antennas would also be corner bent, targeting mobile terminal applications. Even though polycarbonate antennas would also achieve landscape and portrait modes of operation, the substrate suffers from high dielectric loss tangent and hence leading to deterioration in the effective gain in the forward direction. It must also be noted that polycarbonate based antenna design is not an industry ready or industry standard process. The fabrication process flow has to be redone to comply with the polycarbonate. In this chapter, the fabrication process flow is the same as the conventional industry standard PCB manufacturing process and it seems this can easily be manufactured.

The antenna elements have a shared ground with a high gain yield with minimal effective radiating volume. The design principle of shared ground with multiple antenna elements operating for different orthogonal directions is investigated as against the overlapped architecture presented in Chapter 3. It is interesting to note that with a shared ground design strategy, an industry standard fabrication process could be used for the manufacture of the antenna elements as well as the antenna module, with pattern diversity which supports the aforementioned use cases. In previous chapters, all the antennas were characterized in an anechoic chamber, which indicates the antenna's free space behaviour. In order to study the performance of the antennas in an indoor environment, the proposed antenna is also deployed in a real-world indoor environment to study the received power profile.

4.2 CPW-fed Conformal Folded Dipole with Pattern Diversity

It has been demonstrated in multiple reports that the desirable features of antennas integrated on a 5G mobile terminal include conformal structure, hence occupying the least volume to achieve the desired gain, a low specific absorption rate, a high impedance bandwidth to facilitate future bands near 28 GHz, and high gain for the available aperture at the mobile front-end. Pattern diversity is also a desired feature to support the landscape and portrait modes of the smartphone [1–6]. Several designs for antennas to be integrated on 5G mobile terminals have been proposed. For example, the multi-layered antenna array would increase the complexity, and the aperture size might be unsuitable for the mobile terminals [7]. The circularly polarized narrowband high gain element proposed in [8] has an aperture of 25 mm × 25 mm, which exceeds the available space in a mobile terminal. The antenna designed in [9] has 58% wideband with stable patterns and a reasonable gain of 6–8 dBi but suffers from high SAR if integrated on a mobile terminal. The radiation patterns of the strongly resonant structure reported in [10] are unusable for the mobile terminal in any feasible orientation. It must be noted that most of the reported papers have planar design with microstrip feed.

The fabrication process described in [11] to realize a conformal antenna element could be redesigned for the 28 GHz band, but the dielectric mount supporting the radiating element would create an impedance mismatch in addition to reduction in gain.

The conformal antenna array of [12] has a large electrical footprint and hence might lead to distortions in the beam when scaled to the 28 GHz band. The 60 GHz antenna array of [13] has a conformal radius of 25 mm, translating to 2.5λ at 28 GHz, indicating a large physical footprint of the element. A uniplanar feed mechanism would be preferred for a smooth transition in conformal architecture. Conformal designs also lack pattern diversity for integration with mobile terminal.

CPW feed for the antenna is readily feasible for substrates with relatively higher dielectric constant at lower operating frequencies below X-band [14]. The gap width in the CPW line would be plausible with the standard chemical etching method for the aforementioned criteria. Also, the uniplanar feeding technique is popular beyond 60 GHz because of the application of micro-fabrication processes and the probe stations utilized for antenna characterizations [15]. The CPW-fed element proposed in [16] has a wide bandwidth but low gain 1.5–4 dBi. The triple band antenna in [17] has a narrow bandwidth at 30 GHz, and an air bridge is part of the feedline for the impedance match. The considerations for designing CPW transmission lines are explained in [18].

4.2.1 CPW-Fed Folded Dipole

The proposed CPW-fed folded dipole is illustrated in Figure 4.1(a), and its corresponding photograph of the fabricated prototype in Figure 4.1(b). It is designed on a Nelco NY9220 substrate with ε_r of 2.2 and 20 mil thickness. Typically, substrates are characterized for the dielectric constant at 10 GHz, and the manufacturer would specify the tolerance for the substrate. In this case, the dielectric constant varies within 2.2 ± 0.02.

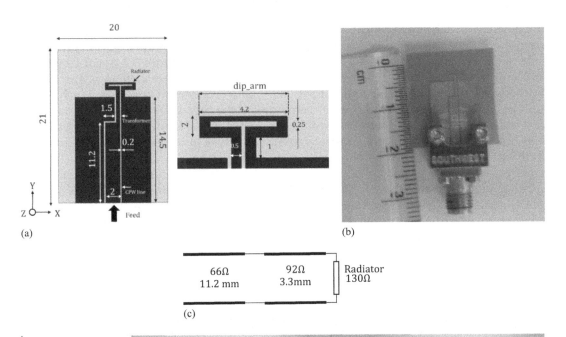

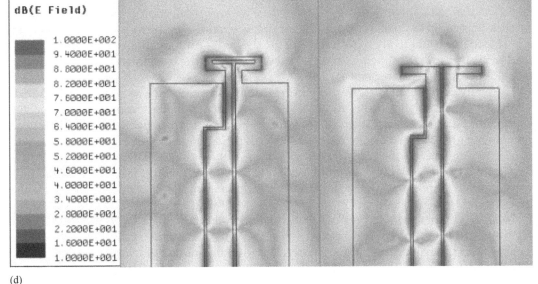

FIGURE 4.1
(a) CPW fed folded dipole design, (b) Photograph of the folded dipole, (c) Equivalent circuit of the antenna, and (d) E-field plots of folded and conventional dipole [40].

It would be a good design practice to measure the dielectric constant of the available substrate in the desired frequency band by making a ring resonator or a standard inset-fed patch antenna. The loss tangent could also be calculated by measuring the transmission loss for a standard 50 Ω line for a 50 mm line of the substrate in a properly calibrated vector network analyser. It is also recommended to measure the thickness of the substrate by using digital Vernier calipers or a digital screw gauge. The thickness of the copper would be 15–35 μm. The thickness of the copper would not affect the performance of the antenna.

The only difference between the choice of substrate for a broadside antenna and an end-fire antenna is that, since the primary radiation is along the end-fire axis, the contribution from the surface waves would be much more dominant in the end-fire than in the broadside radiation. It is also observed that the thickness of the substrate is a major concern in end-fire radiating antennas, especially for microstrip-fed end-fire antennas. Electrically thicker substrate would create additional E-fields between the radiator arms, which would lead to higher cross-polarization levels compared to the broadside counterpart. As a general rule of thumb, a 10 mil substrate would be an ideal choice for designing microstrip-fed end-fire antennas. But a 20 mil substrate would also work at the expense of an additional 5 dB of cross-polarization radiation from the antenna. It would also be evident that the cross-polarization levels observed in the corner bent antenna presented in Chapter 2 has a relatively lower cross-polarization level compared to the ones presented in this chapter. But it has to be noted that when the antenna is being fed by the CPW feeding technique the cross-polarization level is not that significant compared to, let us say, a microstrip-fed log-periodic dipole or Yagi–Uda antenna. A low dielectric constant is selected, to reduce the additional modes of surface waves. An electrically thin substrate of 0.05λ was chosen to minimize cross-pol radiation in the end-fire. The chosen CPW line has a characteristic impedance of 66 Ω; this is connected to a stepped impedance transformer with a high impedance of 92 Ω, which acts as a CPW to CPS transition feeding the input impedance of 130 Ω of the folded dipole. It is a misconception that only 50 Ω feed lines must be used for the design of feeding circuitry to the radiating element. This is not necessarily true: the objective of a good antenna design is to achieve an input reflection coefficient of less than -10 dB. This could happen as long as the input impedance is in the range of 27–95 Ω. Hence a 50 Ω feed line is not really required to get the desired input impedance bandwidth, as the load to the transmission line could be varied by adjusting the transformer geometry and its corresponding dimensions.

The equivalent circuit is shown in Figure 4.1(c). This topology was simulated in ADS to verify the characteristics of the input reflection coefficient of the proposed antenna. The concept of equivalent circuit gives an estimate of the antenna behaviour in the circuit domain but gives an inadequate representation of the antenna model. However, this model aids in understanding the impedance behaviour of the antenna. The aperture and radiation patterns of the antenna cannot be understood with this method. A conventional half-wave printed dipole with CPW feed and appropriate impedance transformer was also investigated.

The E-field plots of folded dipole and conventional dipole at 28 GHz are illustrated in Figure 4.1(d). The length of the dipole arms is 2.85 mm, the impedance transformer creates a narrowband transition from the CPW feed line and the radiating arms hence leading to a 10% impedance bandwidth with poor front-to-back ratio. The radiator is more than 1.5λ away from the feed, and the ground width is chosen to ensure proper grounding contact with the end-launch connector with a reasonable pattern in the E-plane. This is standard design practice in the industry, and for the measurement setup used in this section is a mandatory requirement. If the measurement setup is with a rotating horn antenna and the test antenna is mounted on the test jig, then the electrically long antenna might not be required in the first place. The aperture blockage is not that dominant in the end-fire measurements.

It is observed that when the dipole is below 3 mm, the antenna has a poor impedance match and over 5 mm creates an over-moded antenna for the designed feedline and transformer. If the transformer is redesigned, then the corresponding feeding structure also has to be modified to match the impedance with the radiator. Hence, 4.2 mm is chosen as the length of the dipole arm, whose $|S_{11}|$ is shown in Figure 4.2. The simulations were performed using the industry standard Ansys HFSS. A perfect electric conductor bridge was used for CPW feeding in the simulation. A test simulation with the connector design was also performed, to illustrate the effects of the connector on the antenna. The reflection coefficient was almost invariant, both with and without the connector, but the simulation time increased phenomenally because of the composite structure of electrically large metal with a reasonable electrical size of the antenna. The 10 dB bandwidth is from 24 to 30 GHz. The return loss characteristics are maintained even with a CPW feedline gap of up to 350 μm. Tapered lines and quarter-wave transformers would lead to a relatively narrow band because of the frequency sensitive input impedance at the balanced CPS line of the folded dipole. A 20% wide impedance bandwidth for a mobile terminal is justified to accommodate future 5G bands near 28 GHz; this is a valid design strategy even for existing 4G commercial smartphones. The integrated antenna of the mobile terminal must be able to operate independently of the operator and the geography. The slight frequency detuning of the measured curve could be attributed to the fabrication tolerances, frequency sensitive variation of the dielectric constant, and the alignment with the end-launch connector.

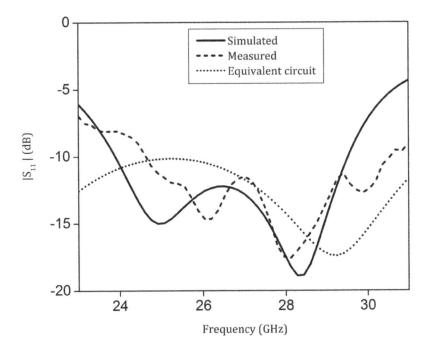

FIGURE 4.2
$|S_{11}|$ of the proposed folded dipole [40].

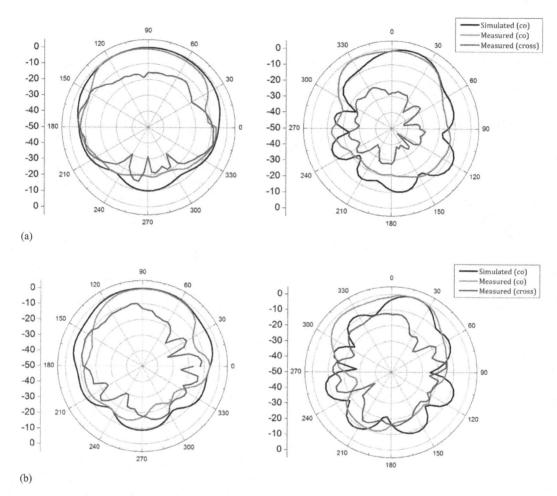

FIGURE 4.3
H- and E-plane patterns at (a) 25 GHz and (b) 28 GHz [40].

The radiation patterns at 25 and 28 GHz are depicted in Figure 4.3. The patterns are stable across the bandwidth because of the uniform mode excitation of the folded dipole. The beamwidth in E-plane (XY plane) is 40°, the pattern is tilted by 15° because the radiator is offset from the phase centre. The narrow beam width is caused by a 1λ wide ground plane, which aids in beam collimation and high front-to-back ratio of more than 10 dB over a 20% band. The beamwidth in the H-plane (YZ Plane) is 140°, with a front-to-back ratio of 10 dB.

The end-fire gain varies from 4 to 6 dBi in the 23–30 GHz band. Gain was measured with the standard gain transfer method. The maximum deviation between simulated and measured gain is 1.6 dB, as illustrated in Figure 4.4. Gain could be increased by 1 dB by increasing the width of the ground plane, which would also increase the physical footprint of the antenna.

The proposed antenna is 2.1λ × 2λ at 28 GHz, and the occupied physical volume is relatively large given the aperture requirements for the end-fire gain. In order to shrink the physical size of the antenna element, an additional matching circuit could be used at the

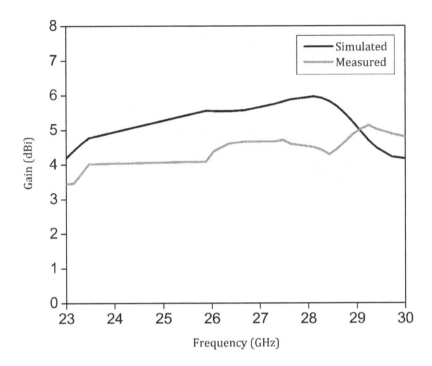

FIGURE 4.4
End-fire gain of the folded dipole [40]. Conformal Folded Dipole

expense of decreased gain. The other alternative is to conform the antenna without a significant compromise in the gain. Hence the proposed topology of conformal antenna does not alter the radiating aperture drastically. The choice of a 20 mil substrate is justified because of the flexibility to conform to the contour of the mobile terminal. Typically, 5–10 mil substrates are the preferred choice but these lack the structural stability for the application in question. It must also be observed that the trace width and gap would be 4.3 mm and 100 μm, respectively, for a 10-mil substrate, leading to an over-moded antenna when matched to the radiating structure. Also, the dielectric support structure for thinner substrates would lead to the detuning of the antenna and create distortions in the H-plane radiation patterns in addition to compromised gain. The uniplanar feed is justified for conformal structures, since a conventional microstrip feed would suffer from higher discontinuity as a result of the bending strain on both sides of the substrate. The 17 μm copper trace on the ground plane would snap when bent, creating a strong impedance mismatch caused by poor transition from the feed to the radiating aperture.

The topology of the conformal folded dipole is illustrated in Figure 4.5. The antenna is bent immediately after the stepped impedance transformer. The primary reason for a 6.5 mm bend is to create the least amount of distortion in the input impedance, as is evident from the bending analysis curves shown in Figure 4.6(a). If the antenna was conformed at the edge of the radiator, the coupling between the orthogonal ground plane and the radiator would detune the antenna element. If the folded dipole is bent including the ground plane, then the back lobe would be enhanced, which could be mitigated with an electrically large aperture of at least 2λ in both dimensions, which might be unsuitable for integration with mobile terminals.

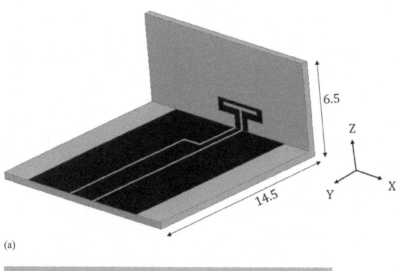

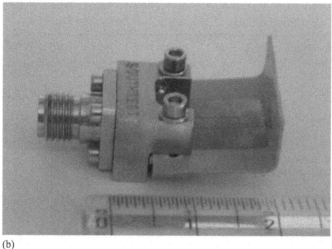

FIGURE 4.5
(a) Design of the conformal folded dipole and (b) Photograph of the conformal folded dipole [40].

The impedance bandwidth of the conformal antenna is from 24 to 30 GHz, as depicted in Figure 4.6(b). It must be observed that the mismatch between planar and conformal return loss characteristics is minimal because of the reduced discontinuity in the feedline post bending. This is caused by the reduced copper footprint on the top plane feeding the radiator. The deviation between simulated and measured results could be attributed to the non-ideal orthogonal bend of the fabricated structure.

H-plane (YZ Plane) radiation patterns are shown in Figure 4.7 at 28 and 30 GHz; beamwidth is around 120° throughout the band. The radiation is a result of the primary folded dipole, and the scattering is caused by the orthogonal feed plane. The beam would be distributed symmetrically in the top half space when the folded dipole is elevated to around 8 mm away from the feed plane. Then the aperture of the reflector would exceed

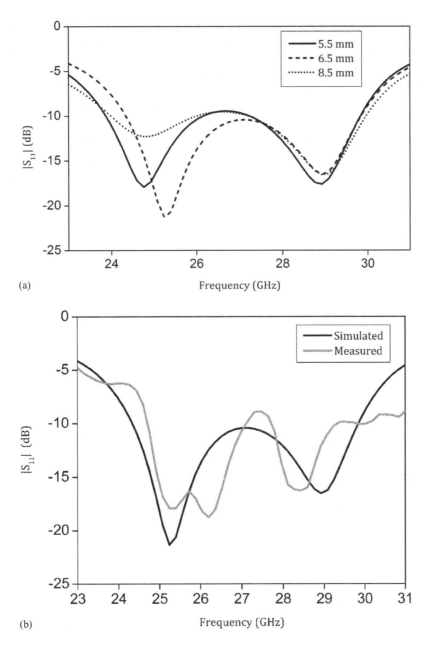

FIGURE 4.6
(a) $|S_{11}|$ variation with bending and (b) $|S_{11}|$ of the conformal folded dipole [40].

the size of the typical mobile terminal for this topology, hence a 6.5 mm height is optimized for reasonable patterns in the H-plane. The radiation in the range 60°–90° indicates that the specific absorption rate would be high when integrated with the mobile terminal. E-Plane (XY Plane) patterns at 28 and 30 GHz are illustrated in Figure 4.7, the beamwidth is 40° with a front-to-back ratio of more than 10 dB across the band. This indicates that the orthogonal ground plane is still effective at maintaining the pattern integrity in the E-plane.

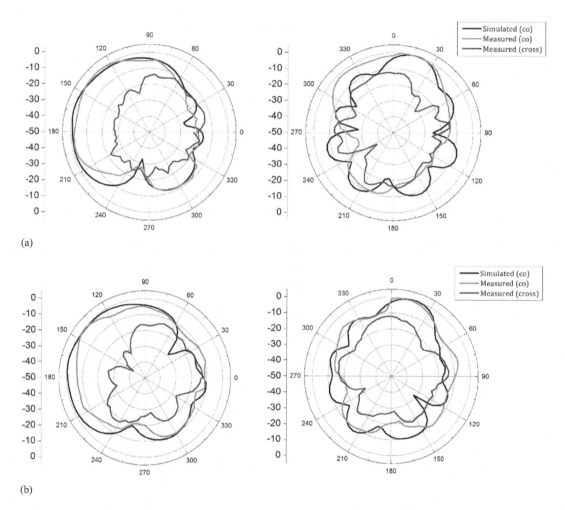

FIGURE 4.7
(a) H- and E-plane patterns at 28 GHz and (b) H- and E-plane patterns at 30 GHz [40].

The beam tilt persists even with the conformal structure because of the offset of the phase centre. The discrepancy between simulated and measured patterns is primarily caused by the transitions utilized for pattern measurements in the anechoic chamber.

The end-fire gain of the conformal antenna is shown in Figure 4.8. It varies from 1.5 to 2 dBi in the 25–30 GHz band. The low gain could be attributed to the reduced effective aperture of the conformal folded dipole antenna compared to its planar counterpart; in particular, the enhanced beamwidth in the H-plane decreased the gain. Various gain enhancement techniques for SAR reduction could be investigated, such as a localized ground plane beneath the folded dipole. This would be operational in a narrow band with poor gain because of the available aperture for conductor backing.

An absorber could be mounted behind the radiating aperture at a quarter-wavelength for back lobe mitigation; this method would reduce the impedance bandwidth and end-fire gain. The third option is to place a wideband reflector strategically near the conformed aperture, which would maintain the 20% bandwidth with a reasonable gain across the band.

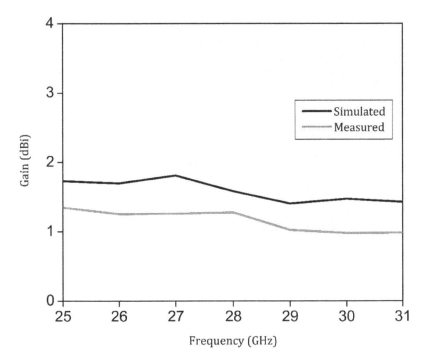

FIGURE 4.8
End-fire gain of conformal folded dipole [40].

4.2.2 Conformal Folded Dipole Backed by Reflector

An electrically large (at least 3λ × 3λ) full metallic sheet could be integrated at a quarter-wavelength near the radiating aperture. But this technique decreases the impedance bandwidth of the conformal antenna, hence a wideband reflector composed of periodic structures is proposed. Also, the wideband reflector would deliver a similar end-fire gain and front-to-back ratio at a relatively smaller aperture, proving its utility in integration with a typical mobile terminal. The proposed unit cell, along with the photograph of the wideband reflector, is shown in Figure 4.9. A sinusoidal slot was etched on the top plane of the Nelco NY9220 substrate with 20 mil thickness. Periodic boundary conditions were used for simulations. The length of the waveguide was optimized for dominant mode excitation, and the polarization of the incident wave was in congruence with the polarization of the folded dipole radiation. The transmission is less than 25 dB with a linear phase across the band of interest proving its utility in reflector application. It is illustrated in Figure 4.10.

An array of 5 × 5 unit cells was designed to function as a reflector. Electrically, the reflector aperture is 0.8λ × 0.8λ, which is relatively compact compared to other reported designs. A 0.5 mm clearance is maintained between the CPW feed plane and the reflector to prevent coupling. The width of the reflector decides the beamwidth in the E-plane: for example, if a 0.5λ wide reflector is used then the beamwidth increases to 105°, thus reducing the gain to 3.2 dBi. The height of the reflector is critical, since the front-to-back ratio must be improved in the H-plane. The schematic and photograph of the proposed antenna integrated with the reflector is shown in Figure 4.11. It must be noted that the reflector dimensions are a compromise between effective radiating aperture and compact size for easier

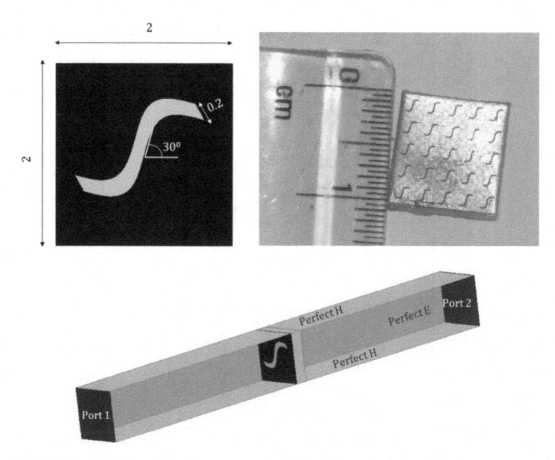

FIGURE 4.9
Unit cell of the wideband reflector along with the simulation model [40].

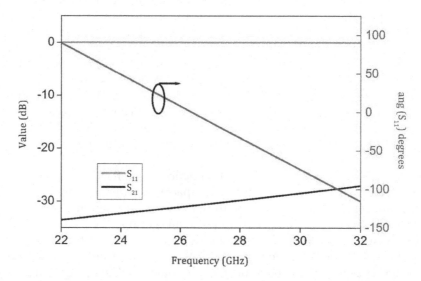

FIGURE 4.10
$|S_{11}|$ and $|S_{21}|$ of the proposed unit cell [40].

Compact Antennas with Pattern Diversity

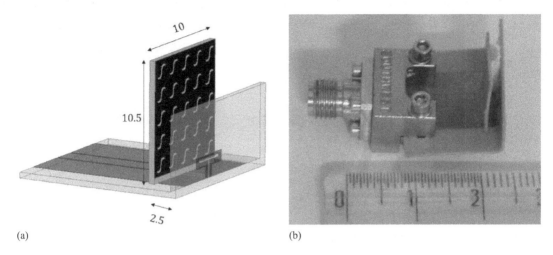

FIGURE 4.11
(a) Conformal antenna backed by the reflector and (b) Photograph of the proposed antenna [40].

integration with a typical smartphone. The offset between the folded dipole and the reflector is the deciding factor of impedance bandwidth and the end-fire gain. To create a volume efficient antenna, the reflector could be placed 0.5 mm away from the folded dipole, but this would reduce the bandwidth and the reflector would behave as a radiator because most of the energy would be coupled to the reflector, where the half-wavelength folded dipole behaves as a parasitic.

If the offset is more than 3 mm, the impedance bandwidth rises up to 25% but the patterns become specular at the higher end of the spectrum. Hence the offset is chosen at 2.5 mm, taking into consideration the trade-off between compactness and the impedance bandwidth. The input reflection coefficient of the proposed antenna integrated with the reflector is depicted in Figure 4.12. The bandwidth is from 24 to 30 GHz, which translates to 20%. The deviation between simulated and measured curves is caused by improper alignment between reflector and folded dipole. Also, the dielectric spacers utilized in the offset space contribute to deviation.

The generalized equivalent circuit of the conformal antenna backed by a reflector is shown in Figure 4.13. Z_{cpw} is the characteristic impedance of the CPW feedline, which must be chosen to reduce over-moding of the antenna and to facilitate the end-launch connector fixture.

The input impedance offered by the primary radiator is Z_{rad}, which must be matched to the CPW feedline by a suitable impedance transformer with characteristic impedance Z_{tnr}. Z_{dis} is the stepped impedance change due to bending, which would be minimal if the transformer feed lines are electrically thin. Z_{ref} is the impedance acting in shunt by the reflector. The distance between the radiator and the reflector decides the input impedance behaviour. A higher impedance bandwidth could be achieved by a larger gap with poor gain.

The H-plane (YZ Plane) patterns are illustrated in Figure 4.14(a) with a beamwidth of 60° as against 120° without the reflector. It must be noted that with the introduction of the wideband reflector, the front-to-back ratio has improved to 12 dB across the band. The peak power is at 160°, which translates to +20° with respect to the horizontal axis of the conformal antenna.

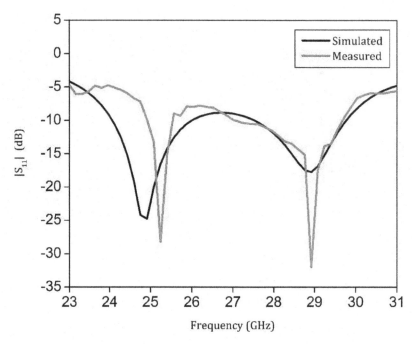

FIGURE 4.12
$|S_{11}|$ of the proposed antenna [40].

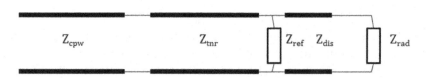

FIGURE 4.13
Generalized equivalent circuit of the proposed element [40].

The power patterns are still suitable for integration with the 5G mobile terminal, since the typical angular tilt of the mobile phone with respect to the user is around +30°, with another 20° tilt. The beam would be directed at +50° towards the base station with at least 10 dB lesser power towards the user for 20% bandwidth. The E-plane (XY Plane) patterns are shown in Figure 4.14(b).

The beamwidth is around 70° with a front-to-back ratio of 10dB. The beamwidth could be further decreased with an increased reflecting aperture. It must be noted that because of the placement of the reflector the beam tilt observed in Figures 4.3 and 4.7(b) has been decreased, resulting from the additional offset by the reflector.

The 3D radiation patterns without and with reflector are depicted in Figure 4.15. It is evident that the H-plane patterns have higher radiation towards the user compared to the E-plane. Hence the height of the reflector is critical for a reasonable reduction in SAR of the integrated mobile terminal. The beam is not uniform because of the scattering effects of the orthogonal feed plane. The proposed topology is an optimal compromise between gain, impedance bandwidth and conformity to standard mobile terminals.

Compact Antennas with Pattern Diversity

Gain in the peak beam is investigated with respect to the offset parameter. The gain has an optimum distance for maximum value. Also, increase in gain corresponds to a decrease in impedance bandwidth. The choice of the offset parameter at 2.5 mm is justified for the stable gain across the 20% bandwidth. Simulated and measured gain curves are shown in Figure 4.16. Gain varies in the range 6–7 dBi in the 25–31 GHz frequency band. The discrepancy between the two curves is a result of the improper mounting of the reflector with the conformal antenna. The maximum deviation is 2 dB between the two curves.

The standard adult human head model from the Ansys Resource Library was utilized to study the SAR values of the proposed antenna elements. It is illustrated in Figure 4.17(a). A cross-section of the human head phantom was used to investigate the SAR performance of the proposed antennas, since the entire human head would be electrically larger than $(20\lambda \times 16\lambda \times 20\lambda)$ which would require high-performance computing for the numerical solution. The SAR values in the cross-section demonstrates the SAR behaviour in the volume. The proposed conformal antennas were mounted near the ear of the model head to

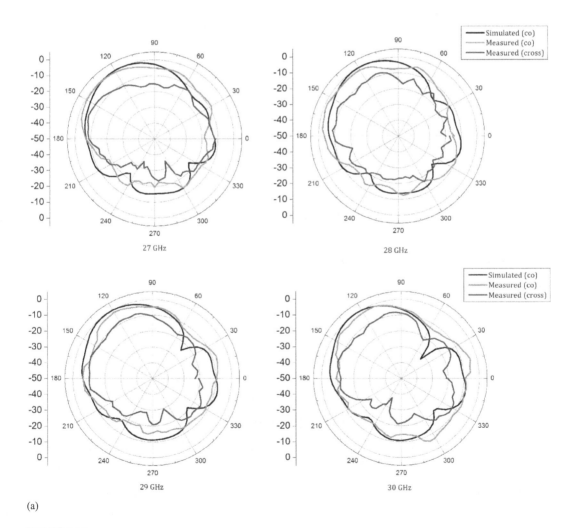

(a)

FIGURE 4.14
(a) H-plane patterns from 27 to 30 GHz, (Continued)

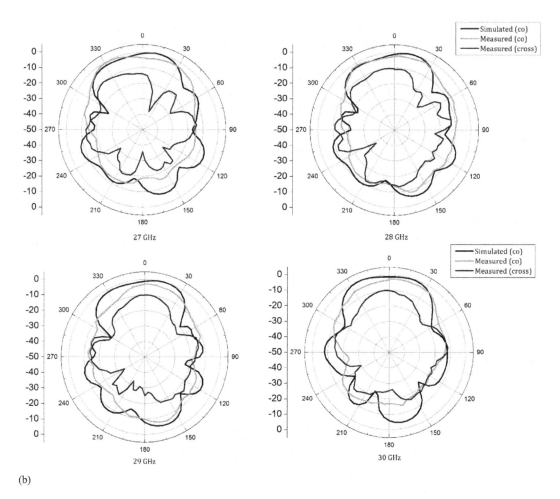

(b)

FIGURE 4.14 (Continued)
(b) E-plane patterns from 27 to 30 GHz [40].

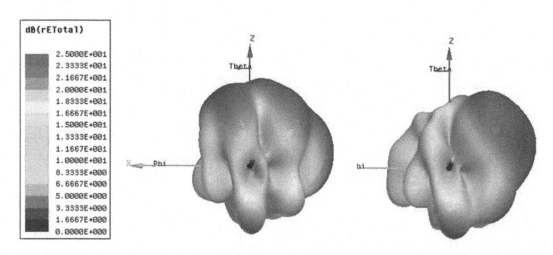

FIGURE 4.15
3D patterns (a) without and (b) with reflector [40].

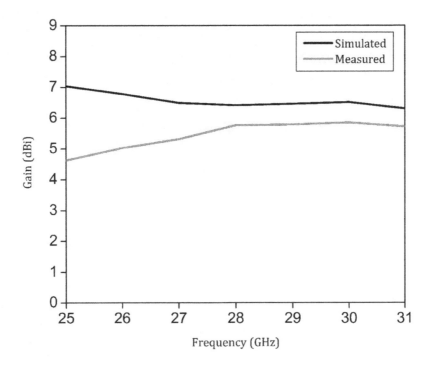

FIGURE 4.16
Forward gain of the proposed antenna [40].

mimic the behaviour of a typical real-world scenario. The SAR values in the cross-section of the model head without and with the reflector is illustrated in Figure 4.17(b). It is evident from the illustration that the reflector is effective in SAR reduction.

Pattern diversity is also investigated for the proposed antenna, which would cater to the landscape and portrait mode in a typical phone. Two identical conformal antennas were modelled on the same substrate. The distance between the two conformal antennas was around 20 mm, and the mutual coupling was less than 30 dB. The mutual coupling increases to 20 dB when the elements are brought close together at 10 mm.

The 3D radiation patterns when the corresponding ports are excited is shown in Figure 4.18. The feed lengths could be reduced for further compaction. The radiation patterns are stable and are almost independent of the other element, justifying the orthogonality of the radiation patterns and proving the proposed antenna's utility in a typical mobile terminal environment.

Table 4.1 illustrates the comparison with the presented design with other reported articles. A CPW-fed planar monopole antenna with an integrated transformer is presented in [19]. The design has two lumped potentiometers, which aid in the gain control of the antenna, even though the design has CPW feeding similar to the end-fire design presented in Section 4.2.2. The primary radiator supports additional modes, leading to a dual beam, which might not be a useful pattern for every use case scenario. Designing this must take in the equivalent circuit models of the variable resistors in addition to the biasing networks as they would also influence the characteristics of the antenna.

A similar dual-beam antenna is also presented in [20], where sub-wavelength metamaterial unit cells are loaded on to the physical aperture of a stepped impedance transformer-fed bow-tie antenna. The stepped impedance transformer is used purely for designing a

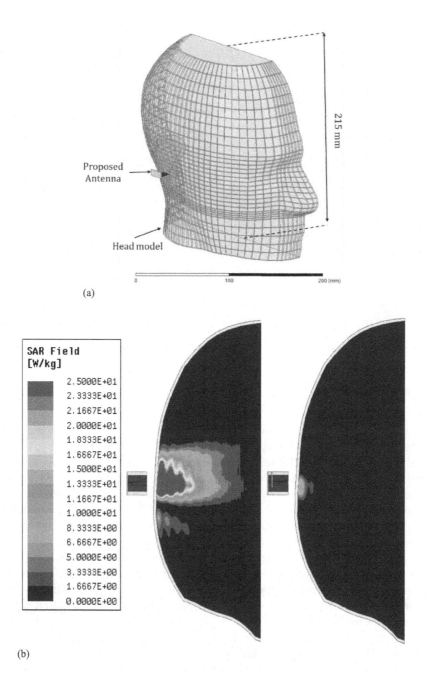

FIGURE 4.17
(a) Antenna placement with head phantom and (b) SAR without and with reflector [40].

non-radiating feed line to the antenna; also, the transformer aids in matching the high impedance of the electrically thin feed line to the bow-tie radiating element. The increase in the gain in the end-fire is close to 2–3 dB.

The application of sub-wavelength unit cells is to change the effective refractive index of the medium of propagation of the waves emanating from the antenna. Additional

Compact Antennas with Pattern Diversity

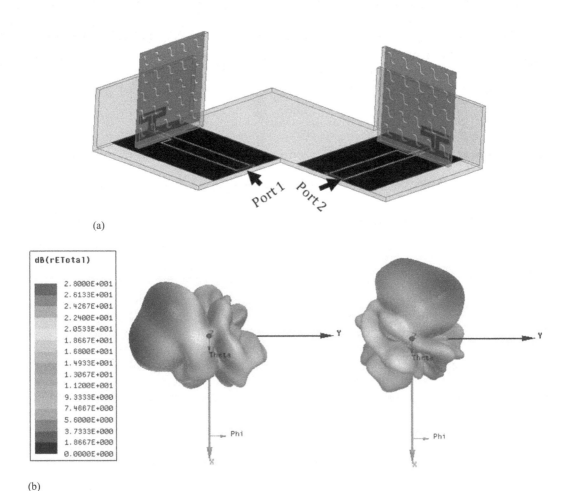

FIGURE 4.18
(a) Pattern diversity design and (b) 3D patterns when Port 1 and Port 2 are excited [40].

TABLE 4.1

Comparison with Reported Articles

Reference	Imp. BW (GHz)	Gain (dBi)	Feed	Conformal
[19]	24–28 (15%)	3–4.5	CPW	No
[20]	57–64 (11.6%)	8–9	Microstrip	No
[21]	22–24 (8.7%)	4–6	Microstrip	No
[22]	21–25 (17%)	8–10	CPS	No
Proposed with reflector	24–30 (20%)	6–7	CPW	Yes

metamaterial unit cells were used for reflecting action. This design strategy might not be applicable for the mobile phone usage scenario, as the antenna is end-fire and the gain enhancement principle used in end-fire would not work in the broadside. Also, this scheme lacks the concept of impedance bandwidth enhancement.

A high efficiency angled dipole is designed at 24 GHz in [21]. The impedance bandwidth is less compared to the presented design. Extension of the same design is reported in [22], where a coplanar stripline (CPS) feeding is used to feed a folded dipole. Typically, the backend circuit feeding the antenna would be either microstrip-based or CPW-based, so an additional baluns or impedance transformer has to be integrated for practical deployment.

4.3 Conformal Antennas with Pattern Diversity

The phased array design proposed in [23] requires precision manufacturing and the scanning loss is almost 2.5 dB when the beam is tilted at 45°; orthogonal pattern diversity is also not a feature of this design. Even though the impedance bandwidth is 36% in [24], beam steering at 90° with respect to bore-sight without deterioration in gain would be challenging to achieve. The 2.6 GHz eight-port array proposed in [25] has a low front-to-back ratio indicating a relatively higher SAR post integration with the mobile terminal. The beam steering designs in [26,27] require high-cost dielectric resonator elements and greater physical footprints. The high radiation efficiency antennas such as a substrate integrated waveguide [28] and a printed ridge gap waveguide [29] need intricate manufacturing processes. The parasitic element switching in [30] also has poor isolation between the states with a greater physical footprint. The pattern diversity architecture proposed in [31] has an electrically compact structure with reasonably high gain, but would occupy higher real estate in an actual mobile terminal. The gain-enhanced, metamaterial integrated multiple-input, multiple-output (MIMO) designs reported in [32,33] are planar, hence yielding lower gain for the occupied volume. Thus corner bent antennas would be an interesting solution for mmWave 5G mobile terminals. Typically, conformal antennas were designed for defence applications, as reported in [12]. The conformal antenna with pattern diversity operating at 60 GHz reported in [13] suffers from gain deterioration post bending of the scaffolding; it must also be noted that the design is electrically large, hence hindering easy integration with a smartphone.

4.3.1 Mobile Terminal Usage Modes

The most popular modes of mobile terminal operation in data mode is illustrated in Figure 4.19. Since the proposed 28 GHz cellular band is intended for data traffic, the antenna design specifications are laid out for this application. Predominantly, the landscape and portrait modes of operation are observed while the user handles the smartphone. Finger blockage might not be very much with the commercial 4G transceiver systems, since the attenuation offered by the human finger is minimal in the 0.7–2.7 GHz band.

However, finger blockage could be detrimental to the communication link at 28 GHz. Several experimental trials of the communication link were performed in the laboratory to demonstrate the attenuation caused by finger blockage. Hence the characteristics of the antennas are as important as their placement in the mobile terminal. The antenna module would be mounted on one edge of the mobile terminal, in a similar way to commercial phones. Statistically speaking, the orientation of the mobile terminal could be either way in the landscape mode, as can be seen in Figure 4.19. Thus finger blockage is critical in

Compact Antennas with Pattern Diversity 93

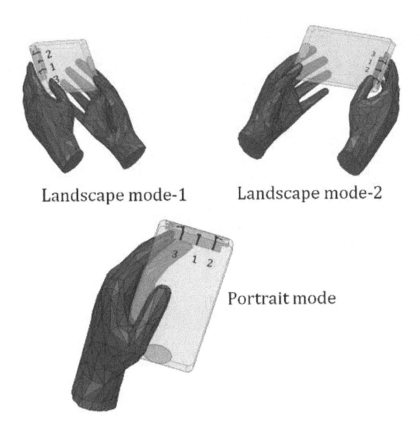

FIGURE 4.19
Various modes of mobile terminal usage.

landscape mode-2. For the proposed configuration in landscape mode-2, the received power decreased by 2 dB at 28 GHz.

Therefore the desired gain of element 3 should be 2 dB higher than that of element 2 to maintain a consistent link budget irrespective of the orientation of the smartphone. The dimensions of the smartphone were chosen from standard literature, which seems to be a statistical average of most commercial smartphones. The length of the prototype smartphone was chosen to accommodate the electrically large end-launch connector. The proposed antennas would operate with a much more compact size and with actual integration with the transceiver module.

Finger blockage is minimal in landscape mode-1 and in portrait mode, as the electrical distance between the finger and the radiating structure is inherently large. The planar MIMO antenna occupies higher volume compared to the conformal or corner bent antennas, as is evident from Figure 4.20, which contrasts planar and conformal design approaches. But the radiation towards the user should be minimal to keep the SAR values within the prescribed limits, and this is a critical requirement from a commercial point of view.

Typically, the average SAR has to be below 1 W/kg for 4G systems, but the standards are yet to be finalized for mmWave 5G systems. Hence a conformal antenna architecture with high pattern integrity and gain compensation is investigated.

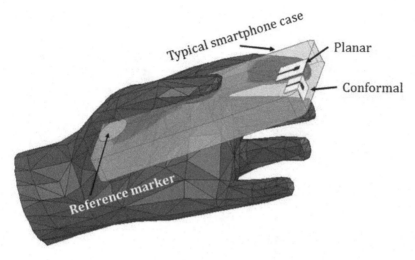

FIGURE 4.20
Comparison between planar and conformal antennas.

4.3.2 Conformal Patch Antenna

The proposed conformal microstrip-fed patch antenna is shown in Figure 4.21(a), and its photograph in Figure 4.21(b). It was developed on a Nelco NY9220 substrate with ε_r of 2.2 and a thickness of 508 μm. A low dielectric constant is preferred, to maintain low excitation of additional surface wave modes. Thicker substrates lead to a 50 Ω line wider than the quarter wavelength at 28 GHz, which would result in additional radiation from the feeding lines, hence decreasing the boresight gain. The feedline of the proposed antenna is 1.2 mm, which has a characteristic impedance of 58 Ω, and it was chosen to achieve a feasible inset width of the radiator. The dimensions of the resonant patch were optimized for an operating frequency of 28 GHz. The overall width of the antenna was decided at 20 mm, to accommodate the 2.92 mm end-launch connector. The planar antenna would have a physically large footprint when integrated on to a smartphone, hence the corner bending of the microstrip patch antenna is investigated.

The transmission line is close to 1λ to reduce coupling between the radiator and the electrically large end-launch connector. The antenna is bent with the radiator, which would be a mere 10 mm that could easily be integrated with most mobile terminal panels. Thinner substrates would be relatively more flexible but would need additional scaffolding, which would alter the radiation characteristics of the antenna, therefore the 20-mil substrate is a reasonable choice.

The 90° bend was carefully implemented after a mild heating of the ground plane of the antenna, since the ground plane would experience higher bending stress compared to the top plane. The bending introduced in the antenna created a few hundred microns of fracture in the ground plane, which was insignificant in the results, hence proving the principle of microstrip conformity in the 28 GHz band.

The impedance bandwidth is from 27 to 29 GHz as shown in the simulated and measured curves of Figure 4.22. The full-wave simulations were done using Ansys HFSS, and all the S-parameter measurements were performed using Agilent PNA E8364C. The input impedance has minimal effect after the 90° bend. To achieve a broad bandwidth, multilayer FSS must be incorporated, leading to a greater physical footprint, thus the bandwidth

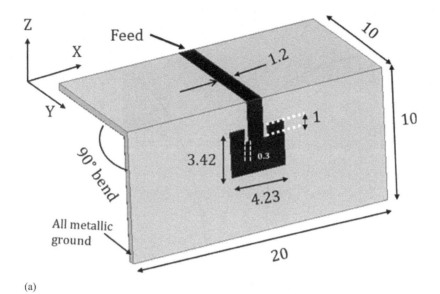

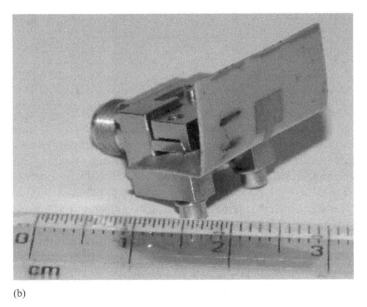

FIGURE 4.21
(a) Schematic of the conformal patch antenna and (b) Photograph of conformal patch antenna.

is compromised in this topology. The discrepancy between simulated and the measured curves could be attributed primarily to the variation in the characteristic impedance of the port assumed in the simulation (50 Ω) and that offered by the end-launch connector utilized during measurements.

The co-polarized and cross-polarized radiation patterns of the conformal patch antenna are depicted in Figure 4.23 for both the principal planes at 27 and 28 GHz. All the pattern measurements were done in the anechoic chamber with a Ka-band Keysight horn antenna as transmitter. The beamwidth in YZ plane (E-plane) is 71° ± 3°, and the beamwidth in the XY plane (H-plane) is 58° ± 3°. The front-to-back ratio is more than 15 dB primarily because

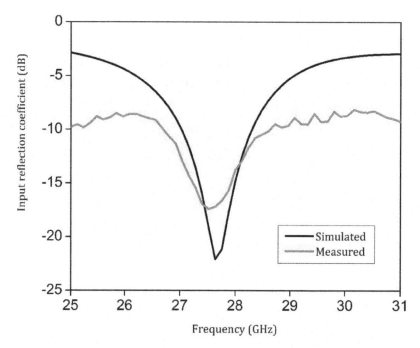

FIGURE 4.22
Reflection coefficient of the conformal patch antenna.

of the electrically large ground plane backing the radiator. The cross-pol radiation is less than 10 dB throughout the band. The patterns of its planar counterpart have similar characteristics. The patterns indicate unidirectional radiation, hence indicating minimal radiation towards the user which in turn means a reduction in SAR values post integration in a mmWave 5G smartphone.

The forward gain of the proposed antenna is 7–9 dBi, as illustrated in Figure 4.24, which is similar to the planar version of the microstrip patch antenna with the same dimensions of the radiator. The measurements were performed using the standard gain transfer method. The discrepancy between simulated and measured results could be attributed to polarization misalignment and frequency sensitive adapters.

The gain yield for the occupied volume is high for the proposed topology. The gain variation is minimal across the band primarily because of the high pattern integrity. Since the transmission line is in the orthogonal plane, the coupling between the feedline and the radiating structure is reduced, leading to higher gain yield. The proposed conformal antenna would be suitable for the portrait mode, as is clear from Figure 4.19.

4.3.3 Conformal Tapered Slot Antenna

Antennas to accommodate the landscape mode-1 of the mobile terminal must be an endfire radiator with a high front-to-back ratio and high gain. The obvious solution to this design is to optimize the mmWave dipole, but the structure would result in a 2–3 dBi gain. Printed log-periodic dipole antennas would lead to gain variation across the band and

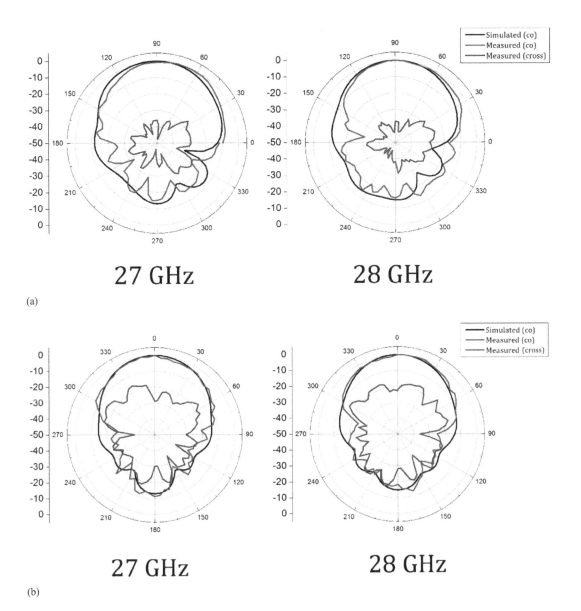

FIGURE 4.23
Patterns in the (a) YZ plane and (b) XY plane.

require multi-stepped impedance transformers for a conformal feed design. Hence tapered slot antenna (TSA) topology is chosen for the application in hand.

The schematic of the proposed antenna is depicted in Figure 4.25(a), and its corresponding fabricated counterpart is shown in Figure 4.25(b). The feed line is a standard 50 Ω, which must be transitioned to the high impedance of 194 Ω created by the microstrip to

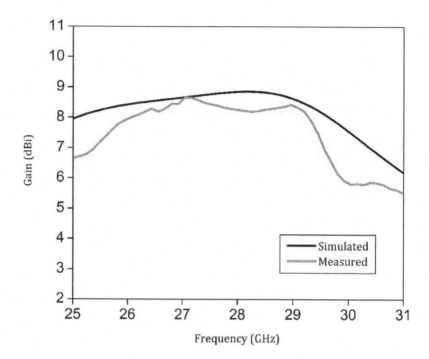

FIGURE 4.24
Forward gain of the proposed antenna.

slotline transition. Hence a quarter-wave impedance transformer of impedance 96 Ω is added in series in addition to the multiple stubs to support the right-angled radiation with respect to the feed. The standard Vivaldi based wideband balun design is avoided here to maintain pattern integrity, resulting in a narrow impedance bandwidth.

The input reflection coefficient of the proposed TSA is shown in Figure 4.26. The impedance bandwidth is 27–30 GHz (10%). The variation in depth of resonance of simulated and measured curves is mainly a result of the solder-free transition from the end-launch connector to the fabricated prototype.

The input impedance behaviour of the conformal antenna is similar to its planar counterpart, indicating a robust design strategy. It must be noted that bending of the 20-mil substrate leads to minimal discontinuities in the ground plane of the antenna.

The simulated and measured radiation patterns are illustrated in Figure 4.27. The beamwidth in the XY plane (H-plane) is 85° ± 5° with a front-to-back ratio of more than 10 dB, indicating low radiation towards the user, when the proposed antenna is integrated with a smartphone. The beamwidth in XZ plane (E-plane) is 55° ± 5°. Since the patterns have a unidirectional wide beam, the link budget could be maintained even with the change in orientation of the mobile terminal with respect to the horizontal axis. The aperture is optimized for a reasonably high gain for the available volume in a mobile terminal environment.

The simulated and measured end-fire gain of the main lobe across the band is illustrated in Figure 4.28. The gain varies from 6 to 8 dBi in the band of 27 to 29 GHz. The gain is 7 dBi at 28 GHz. The gain could have been increased by elongating the physical aperture, but since the objective was to obtain high gain for a compact geometry, the gain was optimized for 7 dBi with reasonable coverage.

Compact Antennas with Pattern Diversity

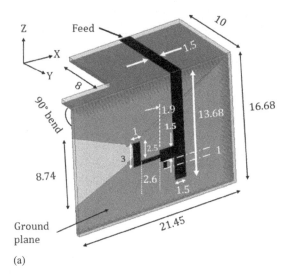

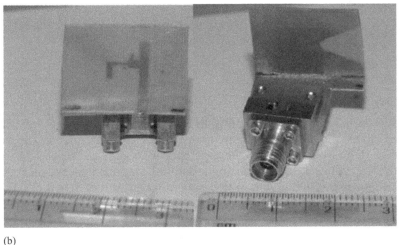

FIGURE 4.25
(a) Schematic of the proposed conformal TSA and (b) Photograph of conformal TSA.

Table 4.2 illustrates the advantages of corner bending with a microstrip feeding technique. Effective radiating volume is the electrical measure of the physical volume, which is responsible for the dominant radiation of the antenna. In other words, it is the effective size of the antenna without the feeding structure of the geometry. As observed in the table, the corner bent topology occupies minimum volume with a higher gain yield, indicating that the module would result in a compact solution to be integrated with mobile terminals.

4.3.4 Conformal TSA with Parasitic Ellipse

In accordance with the modes of operation of a typical smartphone seen in Figure 4.19, the desired gain of the antenna designed for landscape mode-2 must be 2 dB higher than the

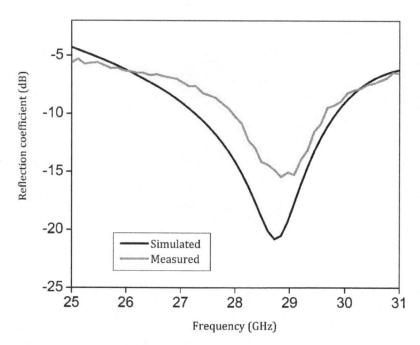

FIGURE 4.26
Reflection coefficient of proposed conformal TSA.

antenna proposed in section 3.3.3, in order to maintain the link budget for both the landscape modes. Hence the tapered slot antenna could be elongated along the axis of radiation, but this strategy would lead to lower aperture efficiency, and to low gain yield for the effective radiating volume. Integrated metamaterial loading is a popular technique for gain enhancement of travelling wave antennas but leads to poor aperture efficiency. Thus a simple solution for gain enhancement is proposed in this section where the tapered slot antenna is integrated with a parasitic ellipse, as illustrated in Figure 4.29. The centre of the ellipse is offset from the microstrip to slotline transition for an optimal E-field distribution in the radiating aperture. The 90° bend is introduced, similar to the antennas presented in the previous sections.

The input impedance is not altered after the introduction of the parasitic ellipse, hence leading to a similar input reflection coefficient as seen in Figure 4.30. The corner bend was introduced carefully in the fabricated structure to prevent a fracture in the copper trace of the ground plane. The impedance bandwidth is 27–30 GHz (10%). The discrepancy between simulated and measured curves is a result of the non-ideal 90° bend in the fabricated prototype.

The simulated and measured radiation patterns in both the orthogonal planes are illustrated in Figure 4.31. The beamwidth in XY plane (H-plane) is 65° ± 5°. The beamwidth in the XZ plane (E-plane) is 33° ± 3°, where the reduction in beamwidth in the E-plane has resulted in a consequent increase in the forward gain by close to 2 dB. The front-to-back ratio is more than 10 dB, primarily because of the electrically large ground.

Compact Antennas with Pattern Diversity

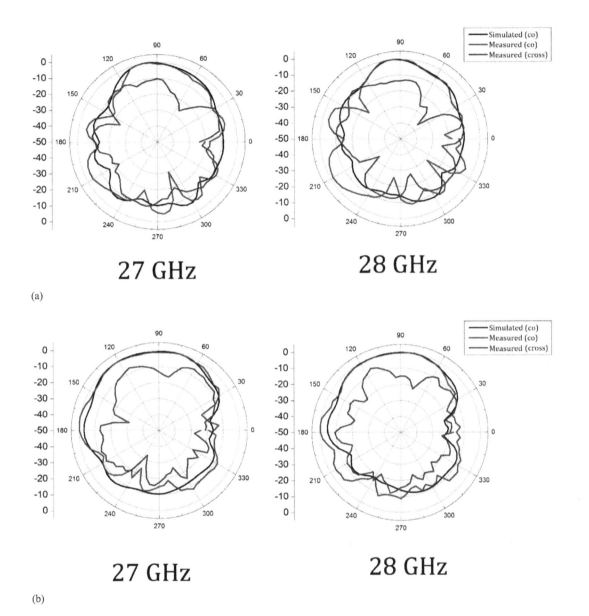

FIGURE 4.27
Patterns at 27 and 28 GHz in (a) XY plane and (b) XZ plane.

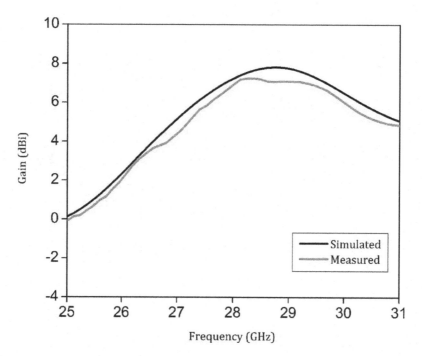

FIGURE 4.28
Gain of the proposed conformal TSA.

TABLE 4.2

Comparison with Reported Articles

Ref.	Freq. (GHz)	Gain (dBi)	ERV*	Corner bent
[20]	60	9	0.144	No
[32]	28	11	0.05	No
[34]	60	12	1.09	No
[35]	64	11	0.08	No
Proposed (patch)	28	9	0.08	Yes
Proposed (TSA)	28	8	0.016	Yes

The simulated and measured forward gains are illustrated in Figure 4.32. The gain is close to 9 dBi, the finger blockage in landscape mode-2 would bring down the effective gain to 7 dBi, consistent with the gain of the conformal tapered slot antenna presented in section 4.3.3. The calculated aperture efficiency is close to 80%.

Table 4.3 illustrates the high aperture efficiency of using a parasitic ellipse in the radiating aperture of the tapered slot antenna. Also, the other reported articles are planar in nature, where the axis of the feedline and that of the radiation are in line. But in the context of the corner bent tapered slot antenna, the axes of the feed line and the radiation are orthogonal as well as in different planes.

Compact Antennas with Pattern Diversity

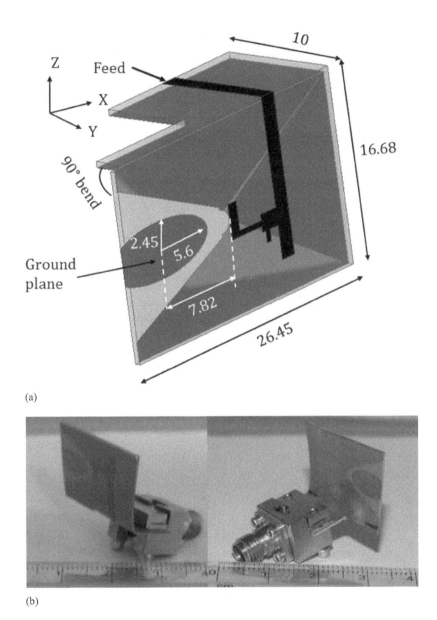

(a)

(b)

FIGURE 4.29
(a) Schematic of TSA with parasitic ellipse and (b) Photograph of TSA with parasitic ellipse.

4.3.5 Conformal Pattern Diversity

All of the three conformal antennas must be integrated to achieve pattern diversity. In order to design a compact module, a shared ground approach is followed. The schematic and the corresponding photograph are depicted in Figure 4.33(a) and (b). The conformal tapered slot antenna would be integrated close to the edge of the smartphone, which experiences the least finger blockage compared to the higher gain element operational for the landscape mode-2. The antennas would be operational even when the port-to-port distance is as close as 0.7λ at 28 GHz, but in order to accommodate the end-launch connectors

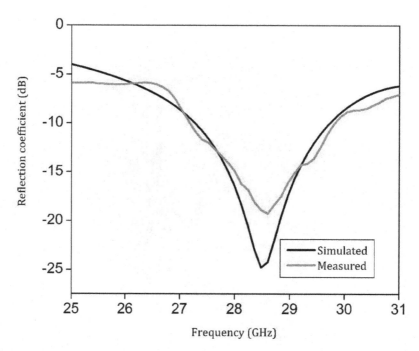

FIGURE 4.30
Reflection coefficient of TSA with parasitic ellipse.

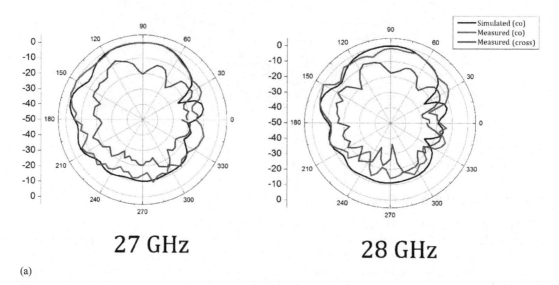

(a)

FIGURE 4.31
Patterns at 27 and 28 GHz in (a) XY plane

(*Continued*)

Compact Antennas with Pattern Diversity

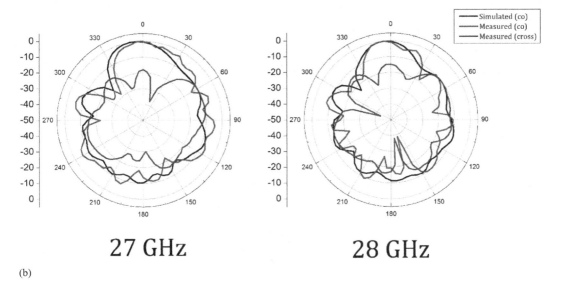

27 GHz   28 GHz

(b)

FIGURE 4.31 (Continued)
(b) XZ plane.

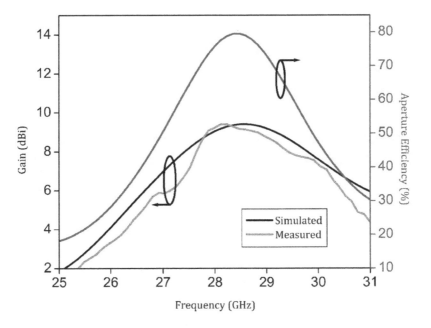

FIGURE 4.32
Gain of the proposed TSA with parasitic ellipse.

for measurements, higher clearance between the ports was maintained, as observed in the photograph. The measured mutual coupling is shown in Figure 4.34.

Mutual coupling is less than 35 dB throughout the band and across the ports. A low mutual coupling is the result of orthogonal radiation from the respective elements. It must also be observed that the patterns post integration are similar to the characteristics of the individual conformal elements, as can be seen in Figure 4.35.

TABLE 4.3

Comparison with Reported Articles

Ref.	Freq. (GHz)	Gain (dBi)	Corner bent	Aperture efficiency (%)
[36]	15	12	No	26.6
[37]	12	11	No	74
[38]	10	14	No	59.5
[39]	61	9	No	63.06
Proposed	**28**	**9**	**Yes**	**80**

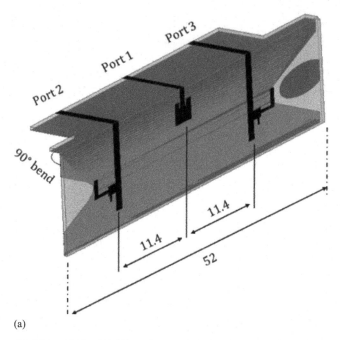

(a)

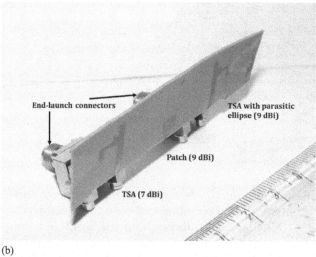

(b)

FIGURE 4.33
(a) Schematic of pattern diversity module and (b) Photograph of the pattern diversity module.

Compact Antennas with Pattern Diversity

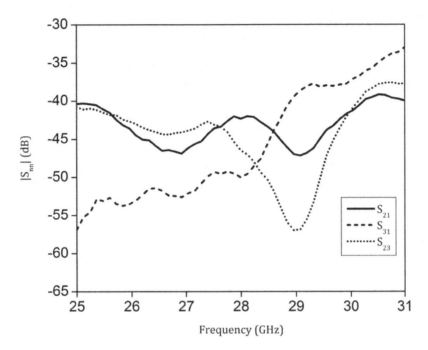

FIGURE 4.34
Patterns of the pattern diversity module with each port excited at 28 GHz.

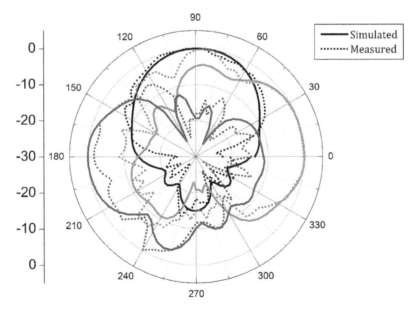

FIGURE 4.35
Patterns of the pattern diversity module with each port excited at 28 GHz.

4.4 Case Studies: Measurement in a Typical Indoor Environment

The laboratory environment schematic is depicted in Figure 4.36. The indoor environment has the typical features of any indoor environment, such as furniture, electric fan, television, etc. The dimensions of the floor are 2 m × 2 m, but the measurement area was chosen to be 1.8 m × 1.8 m for convenience of measurement. The source used was Anritsu MG3694B (up to 40 GHz), the output power was set to +10 dBm, the maximum limit of the source at 28 GHz. The measured cable loss was 26 dB at 28 GHz, excluding the loss caused by the additional adapters. The receiver used was a spectrum analyser Keysight N9010A (up to 44 GHz) and the average measured noise floor was found to be −110 dBm.

The source was connected to the inset-fed patch antenna operating at 28 GHz with a free space gain of 9 dBi. This patch antenna was mounted on the ceiling strategically to illuminate the floor at the centre of the ground. Since the vertical distance between the ceiling and the floor is 3 m, the phase error of the illuminated wave by the patch antenna is negligible across the surface of the ground. Hence the received power is strictly a function of the user's relative orientation with respect to the polarization of the base station antenna and the multipath effects resulting from the environment, such as the metallic cases of the tubelights, wooden furniture, etc.

A 3D printed case with PLA substrate was designed with dimensions 88 mm × 48 mm × 12 mm and a thickness of 2 mm. The conformal patch antenna proposed in Section 4.3.2 was strategically mounted close to the edge of the panel, as seem from the inset of Figure 4.36(c). An all-metallic RF board was also designed and mounted close to the antenna, in a similar way to most of the available commercial smartphones.

Case 1 Mobile Terminal in Portrait Mode

The authors carried out the received power measurements on the grid depicted in Figure 4.36(b). The received power profile in shown in Figure 4.37, which is in the range of −85 to −65 dBm in the coverage area of 1.8 m × 1.8 m. Since the polarization match was relatively higher, the received power was also reasonable to achieve the link budget. It must also be observed that the received power was close to −100 dBm, i.e., 10 dB above the noise floor even at a distance of 6 m from the aperture of the base station antenna. Hence, if separate networks for indoor and outdoor could be deployed, the communication system would work at 28 GHz.

Case 2 Mobile Terminal in Landscape Mode

Identical measurements were used for the landscape mode. The prototype smartphone was held in the orientation that mimics a typical usage scenario. The received power curves illustrated in Figure 4.38 prove that the received power is well above the noise floor.

Compact Antennas with Pattern Diversity

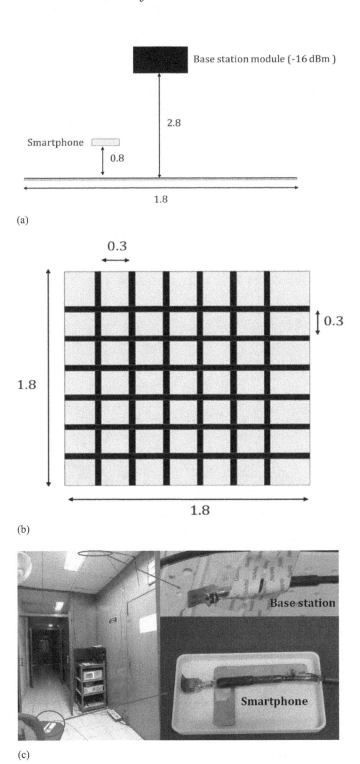

FIGURE 4.36
(a) Side view of the indoor measurement setup, (b) Top view, and (c) Photograph of the indoor environment.

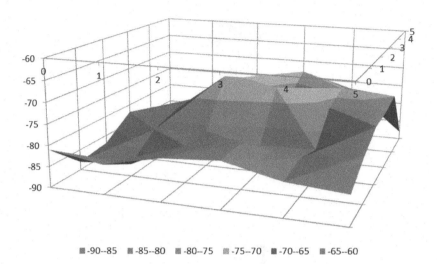

FIGURE 4.37
Received power profile for Case 1.

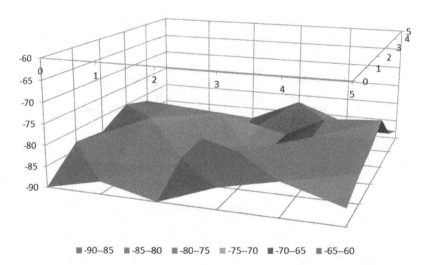

FIGURE 4.38
Received power profile for Case 2.

4.5 Conclusion

Two distinct orthogonal pattern diversity architectures to accommodate landscape and portrait modes were introduced in this chapter. First, a CPW-fed wideband end-fire radiator was introduced and the conformal topology of the proposed antenna was investigated. An FSS reflector was integrated electrically close to the folded dipole for gain enhancement, and the orthogonal pattern diversity design was also investigated. Second, a conformal microstrip-fed pattern diversity module was investigated with antennas yielding high gain for minimal effective radiating volume. The proposed antenna was also deployed in a real-world indoor environment to measure the received power profile on the ground with both landscape and portrait modes.

References

1. Forecast, Cisco VNI, "Cisco Visual Networking Index: Global Mobile Data Traffic Forecast Update 2009–2014," Cisco Public Information, February 9, 2010.
2. C.-X. Wang, et al., "Cellular Architecture and Key Technologies for 5G Wireless Communication Networks," *IEEE Communications Magazine*, vol. 52, no. 2, pp. 122–130, 2014.
3. W. Hong, K.-H. Baek, Y. Lee, Y. Kim, and S.-T. Ko, "Study and Prototyping of Practically Large-Scale mmWave Antenna Systems for 5G Cellular Devices," *IEEE Communications Magazine*, vol. 52, no. 9, pp. 63–69, 2014.
4. T. S. Rappaport, S. Sun, R. Mayzus, H. Zhao, Y. Azar, K. Wang, G. N. Wong, J. K. Schulz, M. Samimi, and F. Gutierrez, "Millimeter Wave Mobile Communications for 5G Cellular: It Will Work!," *IEEE Access*, vol. 1, pp. 335–349, 2013.
5. Zhang, J., X. Ge, Q. Li, M. Guizani, and Y. Zhang, "5G Millimeter-Wave Antenna Array: Design and Challenges," *IEEE Wireless Communications*, vol. 24, no. 2, 106–112, 2017.
6. C. Rowell and E. Y. Lam, "Mobile-Phone Antenna Design," *IEEE Antennas and Propagation Magazine*, vol. 54, no. 4, pp.14–34, 2012.
7. O. M. Haraz, A. Elboushi, S. A. Alshebeili, and A.-R. Sebak, "Dense Dielectric Patch Array Antenna with Improved Radiation Characteristics Using EBG Ground Structure and Dielectric Superstrate for Future 5G Cellular Networks," *IEEE Access*, vol. 2, pp. 909–913, 2014.
8. M. Asaadi and A. Sebak, High-Gain Low-Profile Circularly Polarized Slotted SIW Cavity Antenna for MMW Applications," *IEEE Antennas and Wireless Propagation Letters*, vol. 16, pp. 752–755, 2017.
9. S. F. Jilani and A. Alomainy, "Planar Millimeter-Wave Antenna on Low-Cost Flexible PET Substrate for 5G Applications," in *2016 10th European Conference on Antennas and Propagation (EuCAP)*, Davos, Switzerland, pp. 1–3, IEEE, 2016.
10. J.-S. Park, J.-B. Ko, H.-K. Kwon, B.-S. Kang, B. Park, and D. Kim, "A Tilted Combined Beam Antenna for 5G Communications Using a 28-GHz Band," *IEEE Antennas and Wireless Propagation Letters*, vol. 15, pp. 1685–1688, 2016.
11. K. Sarabandi, J. Oh, L. Pierce, K. Shivakumar, and S. Lingaiah, "Lightweight, Conformal Antennas for Robotic Flapping Flyers," *IEEE Antennas and Propagation Magazine*, vol. 56, no. 6, pp. 29–40, 2014.
12. N. Agnihotri, G. S. Karthikeya, K. Veeramalai, A. Prasanna, and S. S. Siddiq, "Super Wideband Conformal Antenna Array on Cylindrical Surface," in *2016 21st International Conference on Microwave, Radar and Wireless Communications (MIKON)*, Krakow, Poland, pp. 1-4, IEEE, 2016.
13. V. Semkin, F. Ferrero, A. Bisognin, J. Ala-Laurinaho, C. Luxey, F. Devillers, and A. V. Raisanen, "Beam Switching Conformal Antenna Array for mm-Wave Communications," *IEEE Antennas and Wireless Propagation Letters*, vol. 15, pp. 15–31, 2016.
14. L.-M. Si, W. Zhu and H.-J. Sun, "A Compact, Planar, and CPW-Fed Metamaterial-Inspired Dual-Band Antenna," *IEEE Antennas and Wireless Propagation Letters*, vol. 12, pp. 305–308, 2013.
15. S. Raman and G. M. Rebeiz, "94 GHz Slot-Ring Antennas for Monopulse Applications," in *IEEE Antennas and Propagation Society International Symposium*, 1995, Digest, Newport Beach, CA, USA, vol. 1, pp. 722–725, IEEE.
16. G. Zhai, Y. Cheng, Q. Yin, S. Zhu and J. Gao, "Uniplanar Millimeter-Wave Log-Periodic Dipole Array Antenna Fed by Coplanar Waveguide," *International Journal of Antennas and Propagation*, 2013.

17. D. M Elsheakh and M. F. Iskander, "Circularly Polarized Triband Printed Quasi-Yagi Antenna for Millimeter-Wave Applications," *International Journal of Antennas and Propagation*, vol. 2015, pp. 1–9, 2015.
18. R. W. Jackson, "Considerations in the Use of Coplanar Waveguide for Millimeter-Wave Integrated Circuits," *IEEE Transactions on Microwave Theory and Techniques*, vol. 34, no. 12, pp. 1450–1456, 1986.
19. Jilani, S. F., S. M. Abbas, K. P. Esselle, and A. Alomainy, "Millimeter-Wave Frequency Reconfigurable T-Shaped Antenna for 5G Networks," in *2015 IEEE 11th International Conference on Wireless and Mobile Computing, Networking and Communications (WiMob)*, Abu Dhabi, UAE, 2015, pp. 100–102, IEEE.
20. A. Dadgarpour, B. Zarghooni, B. S. Virdee, and T. A. Denidni, "Single End-Fire Antenna for Dual-Beam and Broad Beamwidth Operation at 60 GHz by Artificially Modifying the Permittivity of the Antenna Substrate," *IEEE Transactions on Antennas and Propagation*, vol. 64, no. 9, 4068–4073, 2016.
21. R. A. Alhalabi and G. M. Rebeiz, "High-Efficiency Angled-Dipole Antennas for Millimeter-Wave Phased Array Applications," *IEEE Transactions on Antennas and Propagation*, vol. 56, no. 10, 3136–3142, 2008.
22. R. A. Alhalabi and G. M. Rebeiz, "Differentially-Fed Millimeter-Wave Yagi-Uda Antennas with Folded Dipole Feed," *IEEE Transactions on Antennas and Propagation*, ol. 58, no. 3, 966–969, 2010.
23. B. Yu, K. Yang, C. Sim, and G. Yang, "A Novel 28 GHz Beam Steering Array for 5G Mobile Device with Metallic Casing Application," *IEEE Transactions on Antennas and Propagation*, vol. 66, no. 1, pp. 462–466, Jan. 2018.
24. S. X. Ta, H. Choo, and I. Park, "Broadband Printed-Dipole Antenna and Its Arrays for 5G Applications," *IEEE Antennas and Wireless Propagation Letters*, vol. 16, pp. 2183–2186, 2017.
25. M. Li et al., "Eight-Port Orthogonally Dual-Polarized Antenna Array for 5G Smartphone Applications," *IEEE Transactions on Antennas and Propagation*, vol. 64, no. 9, pp. 3820–3830, Sept. 2016.
26. N. H. Shahadan et al., "Steerable Higher Order Mode Dielectric Resonator Antenna with Parasitic Elements for 5G Applications," *IEEE Access*, vol. 5, pp. 22234–22243, 2017.
27. X. Zhu, J. Zhang, T. Cui, and Z. Zheng, "A Miniaturized Dielectric-Resonator Phased Antenna Array with 3D-Coverage for 5G Mobile Terminals," *2018 IEEE 5G World Forum (5GWF)*, Silicon Valley, CA, 2018, pp. 343–346.
28. W. El-Halwagy, R. Mirzavand, J. Melzer, M. Hossain, and P. Mousavi, "Investigation of Wideband Substrate-Integrated Vertically-Polarized Electric Dipole Antenna and Arrays for mm-Wave 5G Mobile Devices," *IEEE Access*, vol. 6, pp. 2145–2157, 2018.
29. A. Dadgarpour, M. Sharifi Sorkherizi, and A. A. Kishk, "Wideband Low-Loss Magnetoelectric Dipole Antenna for 5G Wireless Network with Gain Enhancement Using Meta Lens and Gap Waveguide Technology Feeding," *IEEE Transactions on Antennas and Propagation*, vol. 64, no. 12, pp. 5094–5101, Dec. 2016.
30. S. Zhang, I. Syrytsin, and G. F. Pedersen, "Compact Beam-Steerable Antenna Array with Two Passive Parasitic Elements for 5G Mobile Terminals at 28 GHz," *IEEE Transactions on Antennas and Propagation*, vol. 66, no. 10, pp. 5193–5203, Oct. 2018.
31. N. Shoaib, S. Shoaib, R. Y. Khattak, I. Shoaib, X. Chen, and A. Perwaiz, "MIMO Antennas for Smart 5G Devices," *IEEE Access*, vol. 6, pp. 77014–77021, 2018.
32. Z. Wani, M. P. Abegaonkar, and S. K. Koul, "Millimeter-Wave Antenna with Wide-Scan Angle Radiation Characteristics for MIMO Applications," *International Journal of RF and Microwave Computer-Aided Engineering*, vol. 29, no. 5, p. e21564.
33. Z. Wani, M. P. Abegaonkar, and S. K. KoulZ. Wani, M. P. Abegaonkar, and S. K. Koul, "A 28-GHz Antenna for 5G MIMO Applications," *Progress in Electromagnetics Research Letters*, vol. 78, pp. 73–79, 2018.
34. A. Dadgarpour, B. Zarghooni, B. S. Virdee, and T. A. Denidni, "One- and Two-Dimensional Beam-Switching Antenna for Millimeter-Wave MIMO Applications," *IEEE Transactions on Antennas and Propagation*, vol. 64, no. 2, pp. 564–573, Feb. 2016.

35. M. Sun, Z. N. Chen, and X. Qing, "Gain Enhancement of 60-GHz Antipodal Tapered Slot Antenna Using Zero-Index Metamaterial," *IEEE Transactions on Antennas and Propagation*, vol. 61, no. 4, pp. 1741–1746, Apr. 2013.
36. I. T. Nassar and T. M. Weller, "A Novel Method for Improving Antipodal Vivaldi Antenna Performance," *IEEE Transactions on Antennas and Propagation*, vol. 63, no. 7, pp. 3321–3324, July 2015.
37. L. Chen, Z. Lei, R. Yang, J. Fan, and X. Shi, "A Broadband Artificial Material for Gain Enhancement of Antipodal Tapered Slot Antenna," *IEEE Transactions on Antennas and Propagation*, vol. 63, no. 1, pp. 395–400, Jan. 2015.
38. B. Zhou, H. Li, X. Zou, and T.-J. Cui, "Broadband and High-Gain Planar Vivaldi Antennas Based on Inhomogeneous Anisotropic Zero-Index Metamaterials," *Progress in Electromagnetics Research*, vol. 120, pp. 235–247, 2011.
39. M. J. Guo, S. Liao, Q. Xue, and S. Xiao, "Planar Aperture Antenna With High Gain and High Aperture Efficiency for 60-GHz Applications," *IEEE Transactions on Antennas and Propagation*, vol. 65, no. 12, pp. 6262–6273, Dec. 2017.
40. G. S. Karthikeya, M. P. Abegaonkar, and S. K. Koul, "CPW Fed Conformal Folded Dipole with Pattern Diversity for 5G Mobile Terminals," *Progress in Electromagnetics Research C*, vol. 87, pp. 199–212, 2018.

5
Pattern Diversity Antennas for Base Stations

5.1 Introduction

The concept of path loss compensation is introduced in this chapter, and two distinct pattern diversity architectures are presented in detail. First, a mmWave tapered slot antenna is introduced, which is integrated with metamaterial unit cells along with dielectric loading. In order to increase gain and its corresponding aperture efficiency, a stacked pattern diversity module is presented to achieve uniform illumination on the ground. Second, spatially modulated sub-wavelength metamaterial unit cells are investigated, which act as phase correcting elements that increase the gain yield for a minimal occupied volume, and the antenna element is used in a co-polarized stacked topology to achieve path loss compensation. The proposed antennas are deployed in a real-world scenario to study the received power profile on the ground illustrating the concepts introduced. The second design illustration is using 3D printed radome, and the third design is a progressive offset metamaterial integrated high aperture efficiency antenna presented along with its corresponding stacked pattern diversity module.

5.2 Pattern Diversity of Path Loss Compensated Antennas for 5G Base Stations

Several reports and articles have argued that high gain antennas are also necessary for base stations [1–3]. One popular approach to the problem is to implement phased arrays on both the mobile terminal and the base station, as demonstrated by researchers at Samsung [4]. This increases the complexity of the mobile terminal because of the intricate switching sequence and locking to the primary beam.

The base station antennas designed to meet the specifications of the existing cellular systems typically have a low insertion loss phased array antenna with fixed beam tilt looking over the ground [5,6]. Similar topology at mmWave frequencies would lead to a poor link budget because of lower gain [7], but an increased gain would result in decreased coverage. Thus it is a trade-off between gain and optimal coverage feasible with reduced complexity to maintain the specified link budget.

Several topologies for 5G base stations operating in the 28 GHz band have been proposed in the literature. For example, the low insertion loss all-metallic tapered slot antenna proposed in [8] has a wide bandwidth of 35% and stable patterns, but the realized design requires precision manufacture. Several research articles have been published on phased arrays, such as the broadband dipole array in [9], even though the mutual coupling is under 20 dB with stable gain in the boresight, gain degrades by up to 3 dB when the beam is oriented at 40°. Hence, phased array architectures would increase the complexity and would be inadequate to provide path loss compensated coverage for the specified area on the ground. The excitation sequence to the phase shifters also increases the complexity.

An alternative approach is to implement a pattern diversity system with a compromise between gain and coverage. The coplanar waveguide (CPW)-fed cross-coupled topology reported in [10] has gain ranging from 4 to 8 dBi and a wide impedance bandwidth. A 360° coverage is expected, but the patterns create blind spots at the higher end of the spectrum. The design demonstrated in [11] has an impedance bandwidth of more than 100% suitable for existing 4G systems, but the beams are not controllable independently.

A third approach is the design of a low-insertion-loss leaky-wave antenna (LWA) such as [12]; even though gain and the side lobe levels meet the specifications of typical base stations, a beam tilt of 30° is observed for 10% variation in the frequency of operation. It must be noted that most of the reported designs do not address the issue of path loss compensation on the ground. Since the path loss is high it is an important design consideration to provide uniform received power for the coverage of interest. The gain of the antenna element proposed in [13] is low because of its electrically compact design. The transmission to the antenna element is through a complicated substrate integrated waveguide, in [14]. The need for high gain antennas is reiterated in [15]. Reflector design for uniform illumination is demonstrated in [16], but the same technique cannot be used for designing a path loss compensated antenna system.

5.2.1 mmWave Tapered Slot Antenna

The standard tapered slot antenna (TSA) is designed on a 0.508 mm thick Nelco NY9220 substrate with a dielectric constant of 2.2 and a loss tangent of 0.0009. Electrically thin substrate was chosen to maintain a low cross-polarization radiation, and the low dielectric constant ensures minimum surface wave modes. The design is depicted in Figure 5.1, with its corresponding photograph in Figure 5.2. Standard radiators such as microstrip patch or

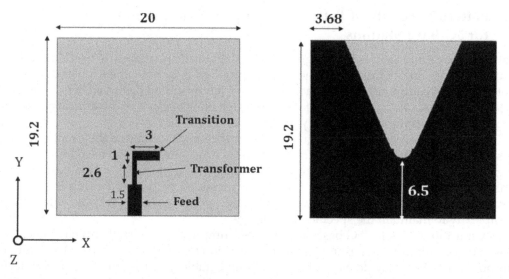

FIGURE 5.1
Top plane and ground plane of TSA central element [29].

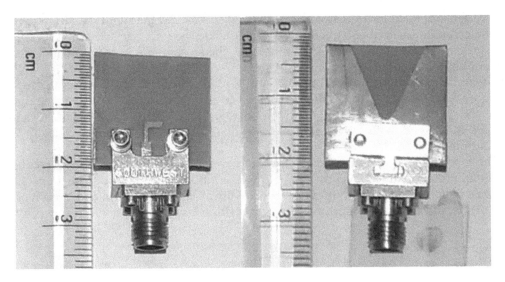

FIGURE 5.2
Photographs of the fabricated TSA [29].

quasi Yagi are avoided here, since the gain bandwidth is relatively less compared to TSA, and the beamwidth cannot be engineered for the application in hand.

The feed line is a standard 50 Ω line in series with a quarter-wave transformer to the 194 Ω microstrip to slotline transition. The flaring angle decides the end-fire pattern, and subsequently the end-fire gain. Standard Vivaldi-based designs usually incorporate a wideband balun, but the pattern integrity deteriorates at the higher end of the spectrum. Base station antennas usually operate with low insertion loss and high aperture efficiency mode. The width of the element is chosen to accommodate the end-launch connector and to achieve a moderately high front-to-back ratio in the end-fire. The tapering and the length of the aperture are chosen to meet the typical specifications of an indoor base station. The input reflection coefficient of the proposed antenna is illustrated in Figure 5.3.

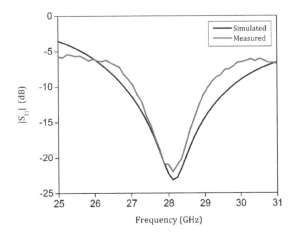

FIGURE 5.3
$|S_{11}|$ of the proposed TSA [29].

The impedance bandwidth is from 27 to 29 GHz (7.2%), and the slight deviation between simulated and measured curves could be attributed to the frequency sensitive variation of the dielectric constant of the low-cost substrate. The measurements were calculated using Agilent PNA E8364C.

The radiation patterns in H-plane (YZ) and E-plane (XY) at 28 and 30 GHz are depicted in Figure 5.4. The beamwidth in the H-plane is 65° ± 5° for the entire band, indicating high pattern stability. The front-to-back ratio is more than 10 dB. Similarly, the beamwidth in the E-plane is 35° ± 5°, with a front-to-back ratio of 13 dB, because of the electrically larger ground plane.

The taper dimensions were chosen for the corresponding beamwidth, which could be mounted orthogonally as a central element of the indoor base station, since it encounters the least path loss. Low transmitter power of 0 dBm at a distance of 3 m and a receiver gain of 0 dBi would require a transmitter gain of 9 dBi to maintain a received power of −62

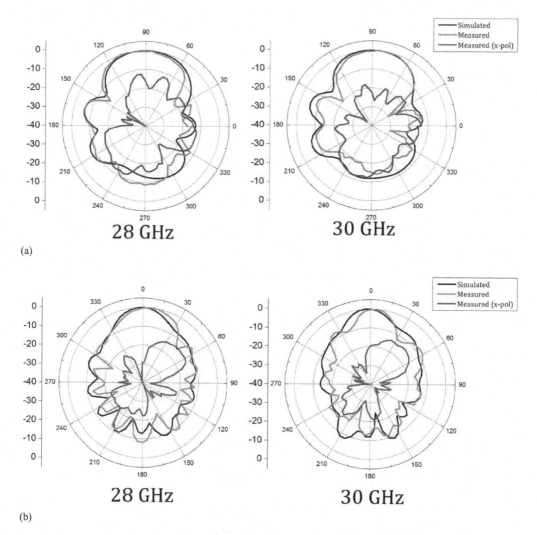

FIGURE 5.4
Radiation patterns at 28 and 30 GHz in (a) H-plane and (b) E-plane [29].

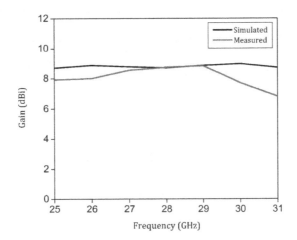

FIGURE 5.5
End-fire gain of the proposed TSA [29]. Application of Metamaterials in mmWave Antennas

dBm, which is a reasonable assumption for the link budget. Hence the aperture has been designed for the aforementioned gain. The aperture efficiency could be further improved by incorporating low loss transitions such as a substrate integrated waveguide (SIW) or printed ridge-gap waveguide feed, but the complexity and manufacturing cost of the design would increase. The simulated and measured end-fire gains are illustrated in Figure 5.5. The gain variation is minimal across the band because of stable patterns arising from the aperture supporting travelling waves, leading to a 1 dB gain bandwidth of 23–32 GHz (32.7%). The deviation between simulated and measured curves is caused primarily by to the utilization of frequency sensitive adapters.

Metamaterials are artificial materials whose properties are not readily available in nature. As an example, negative permittivity and permeability are achieved by using these. In order to design a metamaterial, the sub-wavelength unit cell must be characterized using Floquet ports in the simulator.

In the context of antennas, metamaterials could be used as a superstrate or as a parasitic element in the plane. For a broadside antenna, a metamaterial superstrate could be placed at almost a half-wavelength from the radiating aperture to achieve simultaneous gain and bandwidth enhancement. For an end-fire antenna, an integrated metamaterial loading approach could be followed for gain enhancement. A couple of design examples are presented in the following sections.

5.2.2 Dielectric and Metamaterial Loaded TSA

Since the path loss is of the order of 70 dB at 28 GHz for 3 m distance, it is important to design transceiver antennas that would be able to mitigate the path losses. In an indoor scenario, if a low power transmitter (P_t = 0 dBm) is mounted on the ceiling, then the received power is −62 dBm when the gain of the central element is 9 dBi. When the beam is illuminated at an angle ±45°, the distance increases to 4.24 m, consequently increasing the path loss. In order to compensate for the additional 3 dB path loss, the gains of the antennas illuminating at ±45° must be 12 dBi to achieve a uniform illumination on the ground. Figure 5.6 provides a clear insight into the argument.

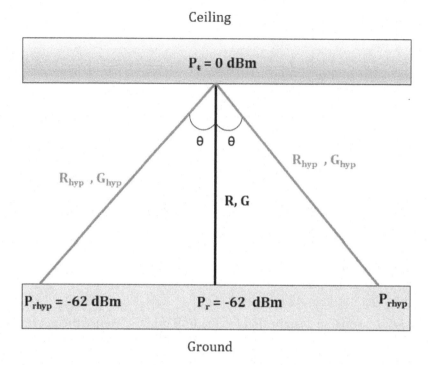

FIGURE 5.6
Gain requirements for the indoor base station [29].

P_{rhyp}, G_{hyp}, and R_{hyp} correspond to received power, transmitter antenna gain and distance, respectively, for the beam angled at 45°. P, G and R are the parameters corresponding to the beam angled at 0°. G_r is receiver antenna gain. In accordance with Friis transmission formula, a uniform illumination would mean $P_{rhyp} = P_r$, thus requiring an additional gain of 20 log(sec θ), as demonstrated in the path loss compensation equations below.

$$P_{rhyp} = P_r \tag{5.1}$$

$$P_{rhyp} = \left(P_t\, G_{hyp}\, G_r\, \lambda^2\right) / \left(4\pi R_{hyp}\right)^2 \tag{5.2}$$

$$G_{hyp} / G = \left(R_{hyp} / R\right)^2 = \sec^2 \theta \tag{5.3}$$

$$G_{hyp}(dBi) = G(dBi) + 20\log(\sec\theta) \tag{5.4}$$

Various architectures could be employed to achieve path loss compensation, such as a phased array approach and cosecant theta pattern synthesis. Typical phased array design at 28 GHz would suffer from a gain deterioration away from the boresight. A 1 × 4 inset-fed patch antenna array with half-wavelength spacing was simulated to characterize the gain degradation effect in the phased array topology. The normalized gain degradation against the angle with respect to the boresight is illustrated in Figure 5.7. It must be observed that a gain drop of almost 3 dB is observed at 45°, hence a phased array needs a gain

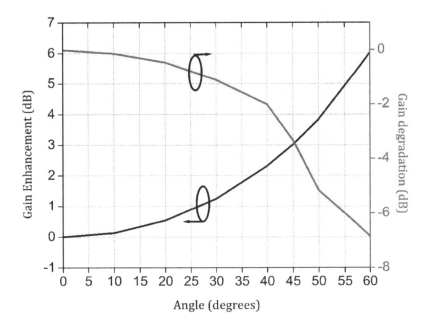

FIGURE 5.7
Gain requirements for path loss compensation [29].

compensation of almost 6 dB at an inclination of 45°. The amount of gain enhancement, GE (dB) = 20 log(sec θ), required with respect to the angle is also illustrated in Figure 5.7. The illustrated gain enhancement is independent of the height of the central element. The second approach to achieve a uniform gain is to synthesize an aperture that would yield a cosecant theta pattern, similar to aircraft tracking radar antennas, but the synthesis of the aperture would be a challenging task, and typically electrically larger apertures lead to high gain and poor coverage. The third approach, the leaky wave antenna suggested above could also be redesigned for uniform gain at the expense of variation in the frequency of operation. Hence an antenna with pattern diversity and compensated gain is the most suitable candidate for path loss compensation with a reasonable coverage.

In order to achieve an end-fire gain of 12 dBi at 28 GHz, the simplest solution is to elongate the aperture of the tapered slot antenna proposed earlier, but the resulting structure would not be a compact solution for the intended base station application. A linear extension of the proposed TSA would lead to a gain of 11.2 dBi. In order to enhance aperture efficiency, the amalgamation of metamaterial unit cells and planar dielectric lens is investigated.

The proposed unit cell is shown in Figure 5.8. The simulation model is also depicted, with the standard periodic boundary conditions and the polarization of the ports matching the polarization of the antenna on to which the unit cells would eventually be loaded. The unit cell consists of multiple angled stubs with sub-wavelength slots. The stubs act as inductors and the slots act as capacitors, hence resulting in a narrowband resonance. The transmission and reflection characteristics are illustrated in Figure 5.9. The transmission is almost 0 dB, indicating a zero attenuation for the incoming cylindrical waves of the antenna. The extracted μ and ε of the unit cell are shown in Figure 5.10, adopted from [17], which indicates a sharp electrical behaviour near 28 GHz, hence if the relative polarization of the incident E-field could be matched correctly with the unit cell, then the E-field could

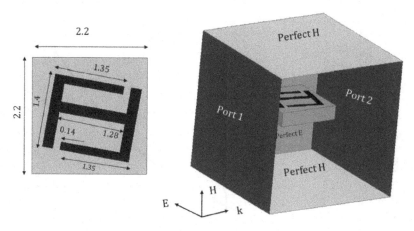

FIGURE 5.8
Proposed unit cell along with the simulation model [29].

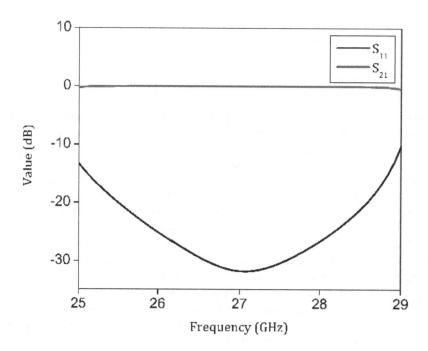

FIGURE 5.9
$|S_{11}|$ and $|S_{21}|$ of the proposed unit cell [29].

be phased corrected by the ensemble of unit cells, hence increasing the gain for the available aperture of the tapered slot antenna. A simple dielectric loading strategy, demonstrated in [18], would lead to lower gain.

The sequential loading of metamaterial unit cells to the antenna element with the corresponding E-fields is illustrated in Figure 5.11. The unit cells on the edge act as reflectors, thus decreasing the coupling between the radiating aperture and metallic ground plane. The metamaterial unit cells operate as localized phase correcting elements. The angular orientation of the unit cells is optimized for phase correction. The phase correction is also

Pattern Diversity Antennas for Base Stations 123

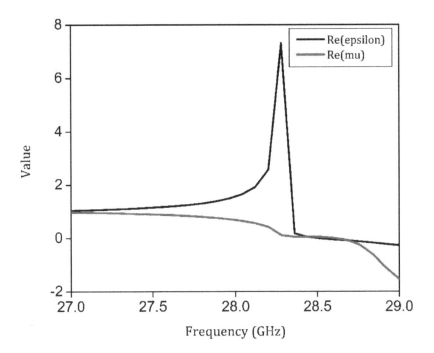

FIGURE 5.10
Extracted permittivity and permeability [29].

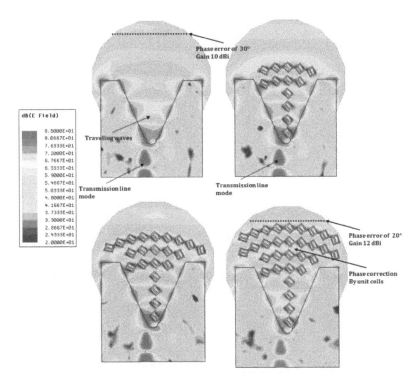

FIGURE 5.11
E-field plots of the design evolution at 28 GHz [29].

evident from the normalized phase against offset plot, as shown in Figure 5.12. The phase error across 1.2λ is 40° for the unloaded element, which is reduced to 20° with the integration of unit cells. The ripples in the phase of the wavefront at the edge of the physical aperture is attributed to the angular separation between the unit cells. The phase linearization behaviour is similar throughout the band.

The proposed element and its photograph are illustrated in Figures 5.13 and 5.14. The dielectric loading of 12 mm radius also creates an effective aperture, hence aiding in gain

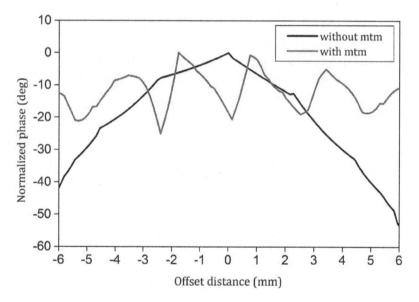

FIGURE 5.12
Phase linearization of the proposed element at 28 GHz [29].

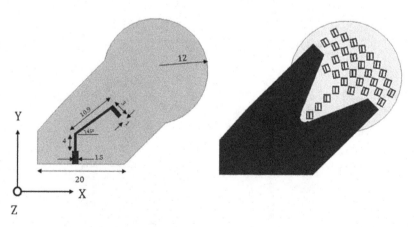

FIGURE 5.13
Proposed TSA with dielectric and mtm loading [29].

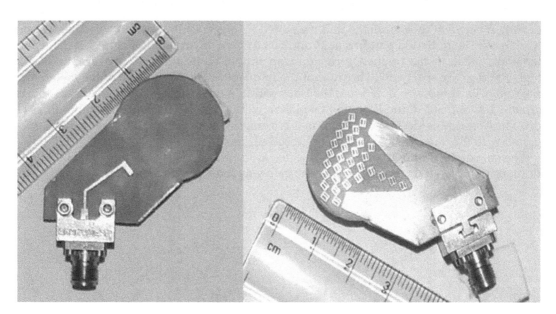

FIGURE 5.14
Photograph of the proposed element [29].

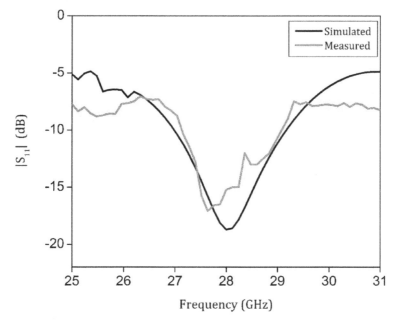

FIGURE 5.15
$|S_{11}|$ of the proposed TSA [29].

enhancement. The aperture efficiency post integration of the metamaterial unit cells is approximately 73%–75%.

The input reflection coefficient of the proposed element is illustrated in Figure 5.15, with an impedance bandwidth of 27–29 GHz. It must be observed that the dielectric and metamaterial loading have a minimal influence on the input impedance of the antenna. The deviation between the simulated and measured results could be attributed to fabrication tolerances. The radiation patterns at 28 and 30 GHz are illustrated in Figure 5.16. The patterns are stable in the frequency range of operation, while the pattern variation at the higher end of the spectrum is because of the frequency-sensitive nature of the metamaterial unit cells. It must be noted that the beamwidth in the H-plane (YZ plane cut at 45°) is around 40°, and 30° in the E-plane (XY plane), with a front-to-back ratio of more than 14 dB. The aperture is designed for optimal coverage when the proposed element is integrated with the tapered slot antenna proposed earlier.

The gain of the proposed element is selected for path loss compensation as per the specifications indicated in Figure 5.7. Hence the gain was chosen to be around 12 dBi at 28 GHz,

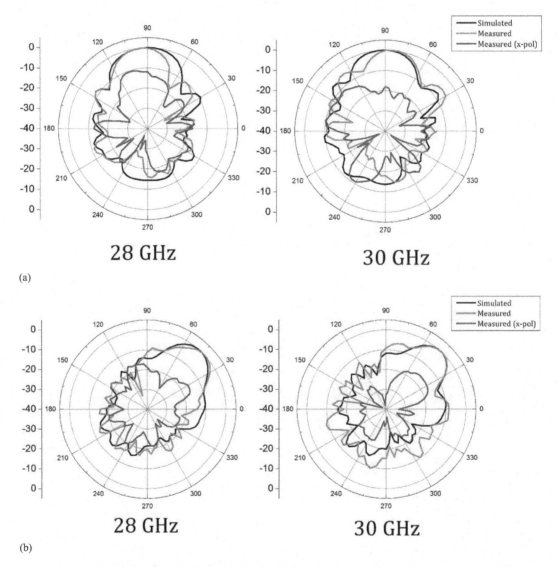

FIGURE 5.16
Radiation patterns at 28 and 30 GHz in (a) H-plane and (b) E-plane [29].

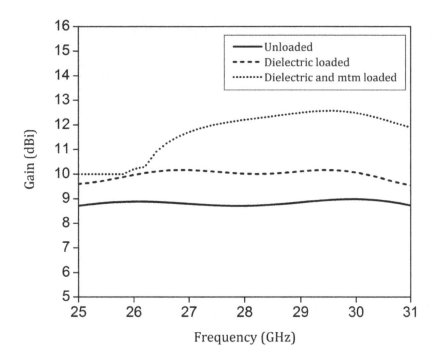

FIGURE 5.17
End-fire gains for different loading of the TSA [29].

with a 1 dB gain bandwidth of 26–32 GHz (20.7%). The gain with subsequent loading of the dielectric and metamaterial unit cells is depicted in Figure 5.17. The dielectric loading alone provides a gain enhancement of 1.5 dB throughout the band, whereas the combination of metamaterial electrical resonators and dielectric loading gives a narrowband gain enhancement of 3 dB, meeting the required specification. Simulated and measured gains are depicted in Figure 5.18, the deviation being a result of the frequency sensitive adapters.

Table 5.1 illustrates the advantages of the proposed concept of metamaterial unit cells and dielectric loading with the tapered slot antenna for gain enhancement. The presented design has high aperture efficiency with high pattern integrity across the bandwidth of consideration. As observed in the table, the non-planar designs yield higher gain as the phase correction of the quasi-cylindrical wavefront emanating from the antenna happens in both of the principal planes of radiation.

5.2.3 Pattern Diversity

A compact topology with stacking for pattern diversity of the proposed elements is proposed here. The architecture for pattern diversity is illustrated in Figure 5.19, and its corresponding fabricated prototype in Figure 5.20. The relatively higher gain elements (12 dBi) directed at ±45° are separated by 22 mm, and the lower gain element (9 dBi) is orthogonally placed between the higher gain antennas. The central element is offset by 10 mm

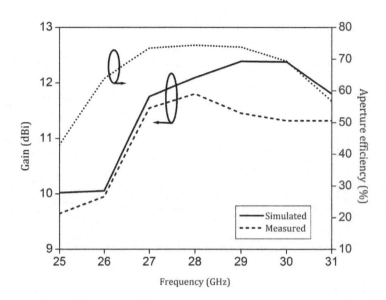

FIGURE 5.18
End-fire gain of the proposed TSA integrated with metamaterials [29].

TABLE 5.1

Design Features of the Metamaterial and Dielectric Loaded Antenna

Ref.	Planar	Freq. (GHz)	Gain (dBi)	1 dB Gain BW (GHz)	Aperture Efficiency (%)
[19]	Yes	15	12	10–20 (66%)	26.6
[20]	No	12	12	12–16 (28.6%)	74
[21]	No	10	14	9.5–12 (23.2%)	59.5
[22]	Yes	61	20.3	58–63 (8.2%)	63.06
[23]	Yes	62	14.5	56–69 (21.2%)	63.6
[24]	No	10	20.7	10–13 (26%)	31
Proposed	Yes	28	12	26–32 (20.7%)	75

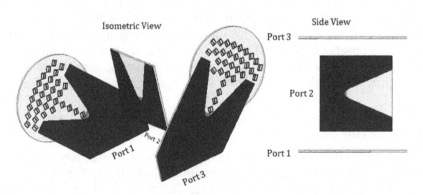

FIGURE 5.19
Architecture for pattern diversity [29].

Pattern Diversity Antennas for Base Stations

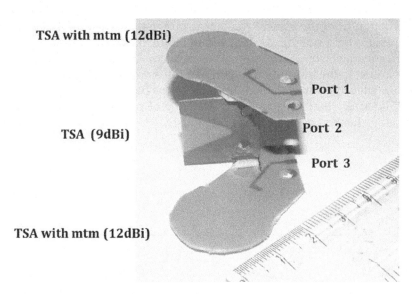

FIGURE 5.20
Photograph of the proposed topology [29].

from the feed plane, as is evident from the side view of the layout, to mitigate the quasi-waveguide effect resulting from electrically large ground planes of the higher gain antennas. Vertical stacking of elements would create blind spots in the coverage angle because of the reduction of the beamwidth of the central element. The radiation pattern of the ensemble at 28 GHz is shown in Figure 5.21. It must also be observed that, since the

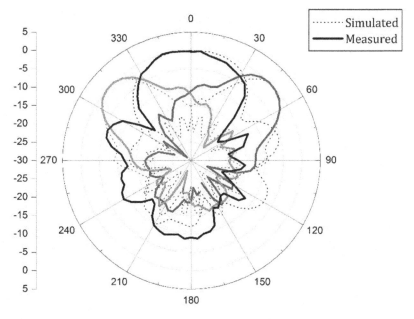

FIGURE 5.21
Patterns when each port is excited at 28 GHz [29].

cross-pol radiation from the central element is minimal as illustrated in Figure 5.4(a), the mutual coupling is below 35 dB, as depicted in Figure 5.22.

A uniform coverage from −60° to +60° is observed. Also, the pattern distortion is minimal in comparison to the individual patterns of the elements in spite of the compact spacing. The beam integrity is maintained throughout the band.

Table 5.2 illustrates that the presented principle of path loss compensation with wide angular coverage has not been reported previously.

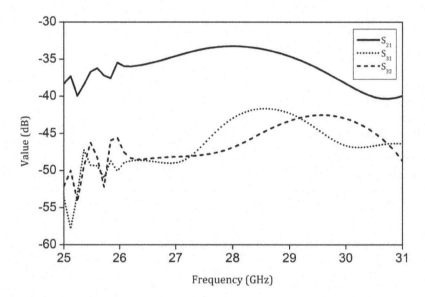

FIGURE 5.22
Mutual coupling between the ports [29].

TABLE 5.2

Comparison of the Beam Switching Module with other Designs

Ref.	Freq. (GHz)	Gain (dBi)	Coverage	Path Loss Compensation
[25]	60	12.4	±35°	No
[26]	60	20	±30°	No
[27]	60	10	±26°	No
[28]	28	11.3	±30°	No
Proposed	**28**	**9,12**	**±60°**	**Yes**

5.3 Path Loss Compensated Pattern Diversity Antennas with 3D Printed Radome

5.3.1 3D Printed Radome for a Patch Antenna

In conventional designs, the antennas lack the radome enclosure. The commercial implementation of the antenna modules has to be well enclosed in a protective casing to prevent damage to the antenna from environmental factors. However, radome designs are usually expensive to implement as the casing must be RF transparent, especially in the 28 GHz band. Hence a simple, low-cost solution is presented.

3D printing is an inexpensive method of realizing any arbitrary geometry within the specifications of the available 3D printer. The chosen material for 3D printing is polylactic acid (PLA) which has a dielectric constant of 2.75 and an associated dielectric loss tangent of 0.01. It is clear that the dielectric is lossy. But arbitrary geometries can be realized in a cost-efficient way, hence the chosen dielectric is justified.

The 3D printed radome made of PLA is placed above the radiator, as radiation studies indicate; its position can be seen in Figure 5.23. It must be noted that the radome enclosure is placed only above the radiating element, as the radiation emanating from the transmission line is negligible. The commercial implementation of the base station modules also has a few millimetres of transmission lines, thus not altering the performance of the radome encased antenna. The thickness of the radome enclosure is 2 mm, which is necessary to maintain mechanical sturdiness. The height between the radiator and the radome is a compromise between the broadside gain and the physical footprint of the module.

Typical all-dielectric superstrates previously reported have a high dielectric constant, thus increasing the cost of manufacture. It must also be noted that conventional substrates might not be compatible with 3D printing, hence making it harder to realize the desired geometry to fit in the antenna module.

The broadside gain variation of the radome enclosed patch antenna for variation of height is illustrated in Figure 5.24. It is clear that the gain deterioration is at a maximum when the radome is placed on top of the antenna without any gap. The gain increases to 11 dBi at 28 GHz when the gap between the radiator and radome is 10 mm. Thus the height

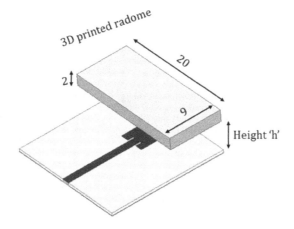

FIGURE 5.23
3D printed radome above the patch antenna.

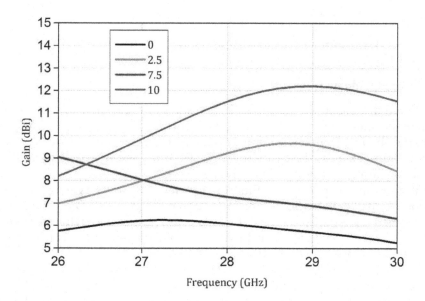

FIGURE 5.24
Broadside gain variation with varying radome height.

of the 3D printed radome could be varied to achieve gain variation, which is necessary to attain path loss compensation.

5.3.2 Pattern Diversity with 3D Printed Radome

The 3D printed radome is shown in Figure 5.25, with the heights of the respective ports optimized to achieve path loss compensation. The beam at 0° would yield a gain of 8 dBi, and the beams at 45° would yield a gain of 11 dBi, hence achieving path loss compensation. It must be noted that the gain is slightly deteriorated because of the enclosure of the sides.

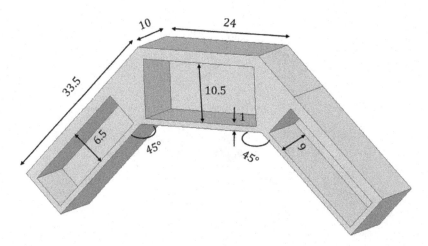

FIGURE 5.25
3D printed radome for the base station.

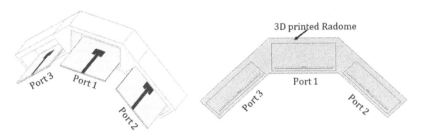

FIGURE 5.26
3D printed radome integrated with the antennas.

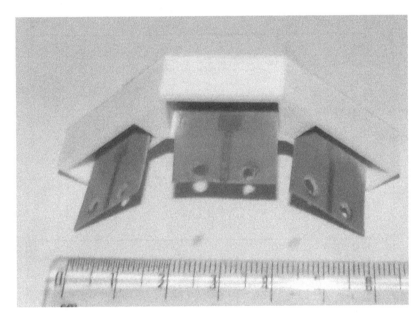

FIGURE 5.27
Photograph of the 3D printed radome with antennas.

The schematic of the pattern diversity module is depicted in Figure 5.26. The photograph of the prototype is illustrated in Figure 5.27.

The input reflection coefficients of the radome-enclosed pattern diversity module is depicted in Figure 5.28. The radome superstrate has a minimal influence on the input impedance of the antenna. The 10 dB impedance bandwidth is 27.5–28.5 GHz. The input impedance of port 2 and port 3 are almost identical. The gains of the mmWave module are shown in Figure 5.29. The gain corresponding to port 1 is 8 dBi, and the gains of ports 2 and 3 are 11 dBi, hence satisfying path loss compensation specifications.

The aperture efficiency corresponding to the antennas of ports 2 and 3 is shown in Figure 5.30; the aperture efficiency being around 72% at 28 GHz. The mutual coupling between the ports is less than 35 dB across the band, as depicted in Figure 5.31. Radiation patterns of the mmWave module at 28 GHz are shown in Figure 5.32. The angular coverage is 140°.

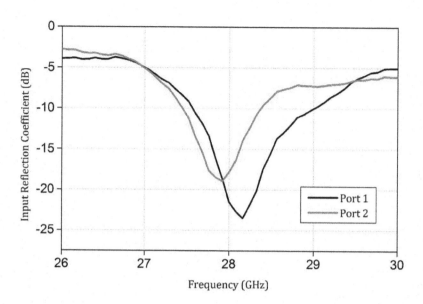

FIGURE 5.28
Input reflection coefficients for the pattern diversity module.

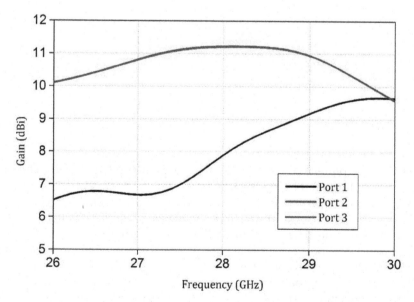

FIGURE 5.29
Gain post integration in the 3D printed radome.

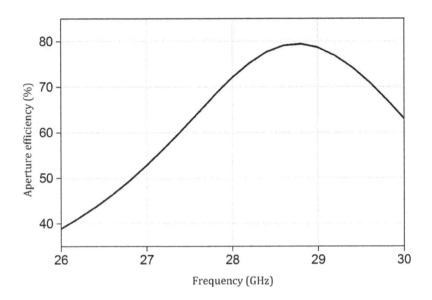

FIGURE 5.30
Aperture efficiency of antennas for ports 2 and 3.

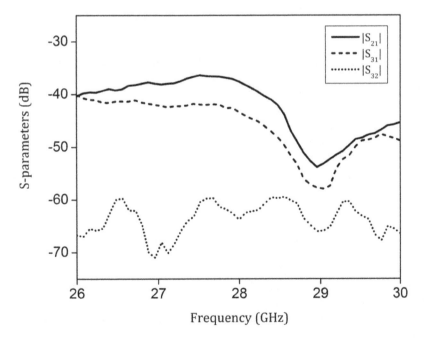

FIGURE 5.31
Mutual coupling between the ports.

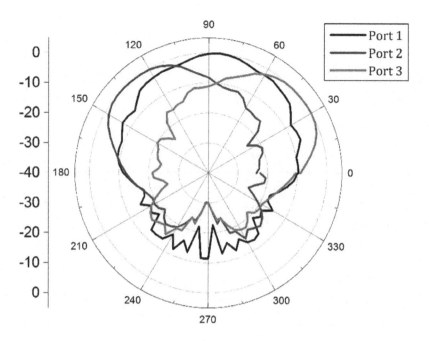

FIGURE 5.32
Radiation patterns of the module at 28 GHz.

5.4 Path Loss Compensated Module with Progressive Offset ZIM

5.4.1 Central Element: Tapered Slot Antenna

The tapered slot antenna with low gain is designed on a 20-mil-thick Nelco NY9220 substrate with a dielectric constant of 2.2 and loss tangent of 0.0009. Electrically thin substrate was chosen to achieve a feasible 50 Ω line with the standard chemical etching process, i.e., 1.5 mm. A thin substrate also ensures relatively lower cross-polarization radiation in the end-fire. The schematic for the tapered slot antenna (TSA) is illustrated in Figure 5.33. The feed line is a 50 Ω line of width 1.5 mm in series with a stepped impedance transformer of 96 Ω of width 0.5 mm connecting to a microstrip to slotline transition of 194 Ω. Conventional variants of Vivaldi antennas would incorporate a radial stub for wide impedance bandwidth, with a compromise in the aperture efficiency. Since the proposed element is intended for 5G base station application, the wideband transition is avoided. The width of the element is 20 mm, 2λ at 28 GHz. The dimension of the conductor width was chosen to ensure the least impedance mismatch between the commercially available 2.92 mm end launch connector and the proposed radiator.

The electrically wide element also aids in achieving a front-to-back ratio of more than 10 dB in the XY plane. The flaring angle of the radiating aperture was chosen as a trade-off between gain and coverage, which would be evident post integration of the element in the base station module.

The simulated impedance bandwidth is 27–29 GHz, translating to 7%. The deviation between simulated and measured graphs could be attributed to a mismatch between the

Pattern Diversity Antennas for Base Stations

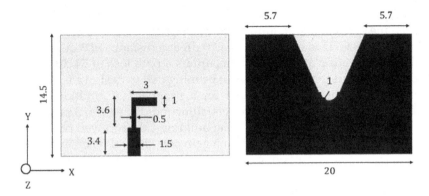

FIGURE 5.33
Top and ground plane of the proposed TSA. (Reprinted with permission from IEEE.)

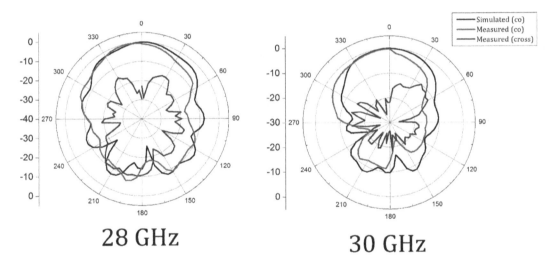

FIGURE 5.34
E-plane radiation patterns of the tapered slot antenna. (Reprinted with permission from IEEE.)

port impedance assumed in the simulation and the actual port impedance of the ports of the vector network analyser (VNA) and the Ka-band connector. The $|S_{11}|$ measurement was performed using Agilent PNA E8364C.

The co-pol and cross-pol radiation patterns of the proposed antenna are illustrated in Figure 5.34 at 28 and 30 GHz frequency. The beamwidth in the YZ plane is 65° ± 5° and 67° ± 8° in the XY plane. The front-to-back ratio is more than 10 dB throughout the band, and in both the principal plane cuts proves its utility in a base station application scenario. The beamwidth in the XY plane is designed for integration with the 5G base station module to achieve an angular coverage of ±65°. The cross-pol radiation is less than 10 dB.

Gain varies between 5.3 and 6 dBi in the 25–31 GHz band. The gain is almost constant because of the high integrity of the radiation patterns across the spectrum. The gain was measured using a standard gain transfer method, with Keysight Ka-band horn antenna utilized as the transmitting antenna. The deviation between simulated and measured curves is primarily a result of polarization alignment error and the use of frequency-sensitive adapters during measurement. The gain choice of 6 dBi at 28 GHz is chosen for angular coverage when the element is radiating in the boresight.

The path loss at 28 GHz is of the order of 70 dB at a distance of 3 m. In contrast to this, commercial WiFi operating at 2.45 GHz encounters a path loss of 54 dB for the same distance. Hence it is evident that path loss must be mitigated to design a feasible communication link. A schematic of the link budget for a proposed 5G indoor base station was illustrated earlier. For a transmitter power of 0 dBm and distance of 3 m, a typical distance between ceiling and floor in most commercial buildings, the received power along the axis is −64 dBm, assuming a transmitting antenna gain of 6 dBi and receiver gain of 0 dBi (also assuming an omnidirectional antenna on the mobile terminal). When the beam is radiated at ±45°, the received power decreases to −67 dBm because of the additional 3 dB path loss. To compensate for the additional path loss, transmitting power could be increased, but the transmitting base stations are usually limited by transmitting power. The antenna gain of the mobile terminal is limited by the constrained physical footprint, hence restricting the available radiating aperture. Therefore the gains of the antennas radiating at ±45° must be 3 dB higher than the boresight antenna element. But the base station module must also maintain coverage without blind spots.

Several approaches could be investigated to achieve the desired path loss compensated gain. Figure 5.35 illustrates the required amount of gain enhancement with respect to the axis, to ensure uniform illumination. The gain enhancement, GE (dB) = 20 log(sec θ), hence a 3 dB additional gain would be required to maintain the same link budget compared to the boresight. The conventional phased array described in various articles suffer from scanning loss and increased sidelobe level, as the beam is tilted away from the primary axis. Leaky wave antenna could be employed for scanning; gain deterioration is observed

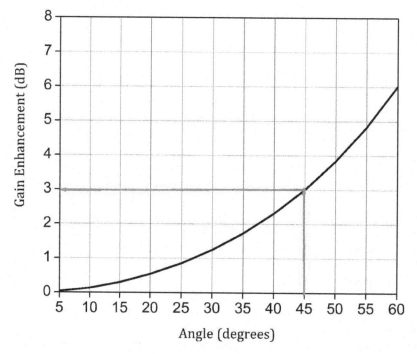

FIGURE 5.35
Gain requirements for path loss compensation. (Reprinted with permission from IEEE.)

as it is scanned with variation in frequency. MIMO antenna designs reported previously offer a compact solution with high gain, but blind spots exist in the patterns of the proposed module. Therefore, pattern diversity with modified gains is investigated.

5.4.2 Spatially Modulated ZIM Loaded Antenna

Several gain enhancement topologies have been reported in the context of Vivaldi antennas. The electrically large parasitic ellipse reported in [19] leads to an increased gain of 2–4 dB beyond 10 GHz, hence leading to low aperture efficiency. It must also be noted that the end-fire patterns become specular at the higher end of the spectrum. A multi-layer zero-index metamaterial integrated antenna presented in [20] achieves a wideband gain enhancement with aperture efficiency of 70%, as a result of phase-correction in both the principal planes. Anisotropic metamaterial in [21] is utilized for wideband gain enhancement in a non-planar design. The gain yield for the physical aperture is low because of the poor phase-correcting characteristics of the proposed unit cell. The metamaterials integrated with Vivaldi based antennas such as [5] also have aperture efficiency of less than 70% across the operational bandwidth.

To achieve an end-fire gain of 9 dBi, the radiating aperture of the element proposed in Section 5.4.1 could be extended from 14.5 mm to 19.2 mm, but in this case the aperture efficiency would be lower. Hence a spatially modulated zero-index metamaterial (ZIM) integrated antenna is proposed, which yields a high gain for the available physical aperture. The unit cell consists of two offset slots which aid in creating a capacitive effect.

The arc-shaped stubs create an inductive effect, hence giving rise to a zero index near the 28 GHz band. The $|S_{11}|$ is less than −25 dB, indicating low reflection, and $|S_{21}|$ is close to 0 dB, indicating a low insertion loss for the incoming quasi-cylindrical waves post integration with the radiating aperture. The E-field was polarized along the X direction and the wave propagation along the Y direction, and periodic boundary conditions for the unit cell simulation were utilized. The relative angular orientation of the slots of the unit cells with respect to the incoming wavefront would decide the amount of phase error correction indicating gain enhancement.

The phase error correction is evident at the edge of the physical aperture. The angle of slot 1 of the unit cell is decided by the angular offset relative to the radiation axis of the antenna. The angular offset of the ZIM unit cells is 0° on the axis and the offset increases progressively as the unit cell is moved away from the axis. Slot 2 has 0° offset for the entire ensemble of unit cells, since the E-field exits through slot 2, hence leading to a phase correction of the wavefront. Since the unit cells are sub-wavelength size, radial loading of these unit cells is necessary to obtain an observable gain. The behaviour is similar throughout the impedance bandwidth.

The phase error is up to 50° for the proposed tapered slot antenna with an end-fire gain of 6 dBi. The phase error is reduced to 15° across 0.8λ at 28 GHz. The ripples observed in the phase distribution reported in previously published articles is also reduced in the proposed topology of spatially modulated ZIM (SMZIM). Because of the reduced phase error of the wavefront, the gain yield for the available radiating aperture is high, also indicating a high aperture efficiency.

The schematic of the proposed antenna integrated with SMZIM is illustrated in Figure 5.36(a), while Figure 5.36(b) depicts the inset view of the unit cells with progressive angular slot variation. The integration of unit cells was performed on the TSA proposed in section 5.4.1. The dielectric shaping of the radiating aperture is to improve the pattern integrity of the element.

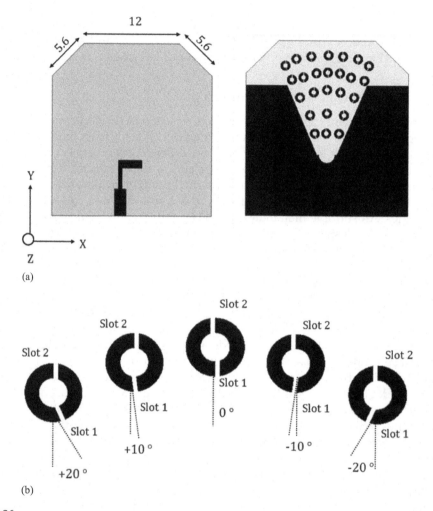

FIGURE 5.36
(a) Schematic of proposed SMZIM integrated TSA and (b) Inset view of the SMZIM unit cells. (Reprinted with permission from IEEE.)

The impedance bandwidth is 27–30 GHz. The SMZIM loading has minimal effect on the input impedance of the element, varying only the radiating aperture. The deviation between simulated and measured curves is caused by alignment errors and fabrication tolerances.

The co-pol and cross-pol radiation patterns at 28 and 30 GHz are illustrated in Figure 5.37. The beamwidth in the YZ plane is 55° ± 2°, and 40° ± 5° in the XY plane. The patterns have high-beam integrity, proving its utility in 5G base stations. Since the beam widths have minimal variation across the band, this element could be integrated in the pattern diversity module. The cross-pol radiation is less than 15 dB in the end-fire. The beam width in the XY plane is a compromise between gain and angular coverage post integration in the 5G base station module. The gain is 8.5 dBi at 28 GHz, which is 3 dB higher than the unloaded TSA. The aperture efficiency is also shown in Figure 5.30. The aperture efficiency is 77% at 28 GHz. The phase correcting elements yields high aperture efficiency. The aperture efficiency bandwidth is narrow because of the angular orientations of the SMZIM unit

Pattern Diversity Antennas for Base Stations

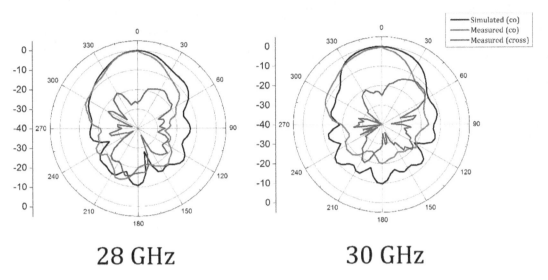

FIGURE 5.37
E-plane radiation patterns of SMZIM loaded antenna. (Reprinted with permission from IEEE.)

cells optimized specifically for 28 GHz. The gain could be further increased by increasing the physical footprint of the antenna, leading to a compromise in aperture efficiency and compactness.

5.4.3 Stacked Pattern Diversity

A compact co-polarized stacking topology is depicted in Figure 5.38. The central element is the standard tapered slot antenna with a gain of 5.5 dBi. The elements angled at ±45° are the antennas with spatially modulated ZIM unit cells with a gain of 8.5 dBi. The elements are stacked vertically for a co-polarized compact topology. The distances between the elements are optimized for clearance for the end-launch connectors. The vertical stacking topology would lead to the quasi-waveguide effect because of the electrically large ground

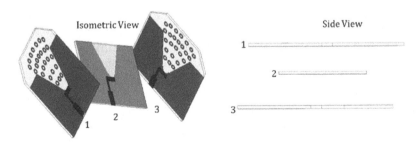

FIGURE 5.38
Architecture for pattern diversity. (Reprinted with permission from IEEE.)

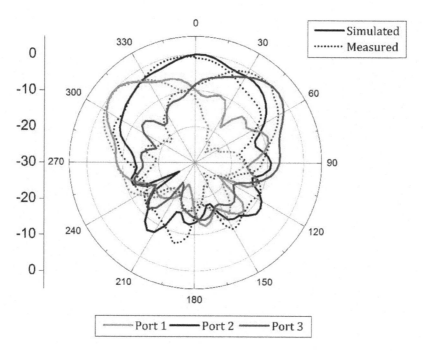

FIGURE 5.39
E-plane patterns when each port is excited at 28 GHz. (Reprinted with permission from IEEE.)

of the tilted elements, thus the central element is offset by 5 mm from the reference planes of the other two elements.

The radiation patterns of the proposed module are shown in Figure 5.39. The coverage is ±65° with path loss compensated gains. The mutual coupling is less than 30 dB at 28 GHz.

5.5 Path Loss Compensated Quasi-Reflector Module

The schematic for the compact pattern diversity module along with the fabricated prototype is illustrated in Figure 5.40. The central element is a CP- fed wideband folded dipole with an impedance bandwidth from 24 to 30 GHz and a forward gain of 6 dBi at 28 GHz, introduced by the authors in section 4.2.1. The antenna proposed in section 5.2.2 is oriented at ±45°. The central element, when it is offset by 4 mm from the reference plane, the electrically large ground planes of the tapered slot antennas create a quasi-reflector effect leading to an increase in the front-to-back ratio, a decrease in the beamwidth and a consequent increase in gain. A 3D printed scaffolding with PLA substrate was also designed to mount the antenna elements. It is evident from Figure 5.41 that the module has a wide angle coverage of ±65° without blind spots at 28 GHz. The mutual coupling is less than 38 dB across the band. The post-integrated gains of the respective elements are shown in Figure 5.42. The gain of the central element has increased to 7 dBi and the gains of the elements at ±45° is 10 dBi, hence achieving path loss compensation.

Pattern Diversity Antennas for Base Stations 143

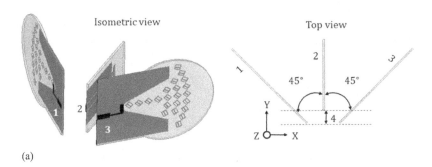

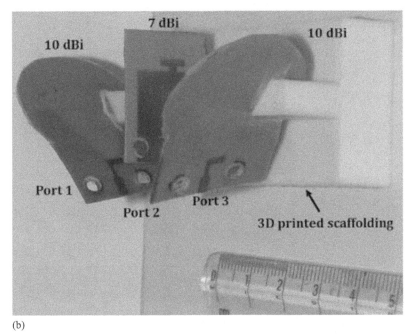

FIGURE 5.40
(a) Schematic of the proposed pattern diversity and (b) Photograph of the fabricated prototype. (Reprinted with permission from IEEE.)

5.6 Design Guidelines for High Aperture Efficiency Antenna

1. Since base station antennas must have both high radiation efficiency and aperture efficiency, the choice of substrate is equally critical. For the design of a tapered slot antenna, the standard 50 Ω line could be used as a feedline to create minimal impedance discontinuity between the end-launch connector and the antenna. It must be observed that electrically thin, low dielectric constant substrates yield a 50 Ω line which has an electrical width of less than 0.1λ, thus preventing radiation from the feedline. This feedline must be transitioned to the slotline using a suitable impedance transformer depending on the desired impedance bandwidth.

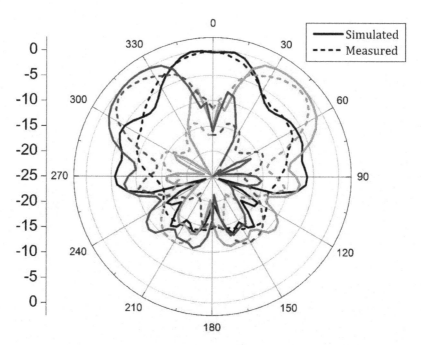

FIGURE 5.41
Radiation patterns of the pattern diversity module at 28 GHz. (Reprinted with permission from IEEE.)

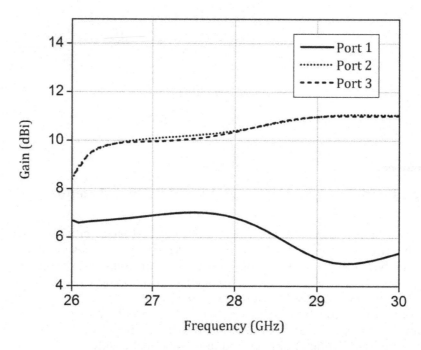

FIGURE 5.42
Gains of the antennas in the pattern diversity module. (Reprinted with permission from IEEE.)

Pattern Diversity Antennas for Base Stations

2. In order to increase aperture efficiency, phase correction must be accomplished. Hence electrical sub-wavelength metamaterial unit cells must be designed that have the least insertion loss ($|S_{21}| \sim 0$ dB) and $|S_{11}| < -10$ dB. Another important design metric is that the sub-wavelength unit cells must create a phase shift of 3°–5°.
3. The arrangement of the metamaterial unit cells is crucial for high gain yield. The phase correction of the wavefront must be analysed carefully by investigating the E-field plots and its localized phase correction.

5.7 Case Studies: Measurement in a Typical Indoor Environment

To study the base station antennas proposed in section 4.3.2, a conformal patch antenna was used in the portrait mode as a receiver connected to the spectrum analyser.

Case 1: Tapered Slot Antenna as Base Station Antenna

The tapered slot antenna (TSA) proposed in Section 5.2.1 was mounted as the transmitting antenna, and the received power profile is depicted in Figure 5.43. Even though the free space gain of the element was 7 dBi, the received power is relatively low compared to the measurements presented in Section 3.4, because of the additional 90° bend transition.

Case 2: Tapered Slot Antenna Integrated with Metamaterials as Base Station Antenna

The high aperture efficiency element proposed in Section 5.2.2 was used as the transmitting antenna, and the conformal patch was used as the receiving antenna in portrait mode. The received power is higher at the edges of the ground, since the element is tilted at 45°. It must be noted that the path loss compensation would be valid in free space with fewer interfering elements from the indoor environment. Figure 5.44. illustrates the received power profile for Case 2.

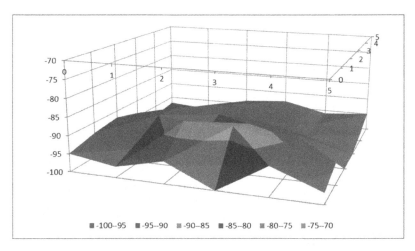

FIGURE 5.43
Received power profile for Case 1.

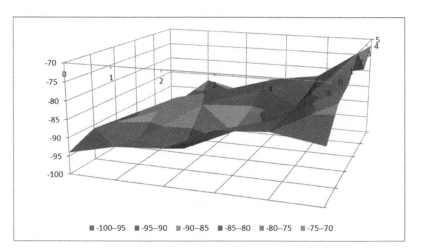

FIGURE 5.44
Received power profile for Case 2.

5.8 Conclusion

The idea of path loss compensation was introduced in this chapter. First, a tapered slot antenna integrated with dielectric loading and metamaterial unit cells was explored for enhancement in gain and aperture efficiency, and a pattern diversity scheme was also presented. Second, a spatially modulated metamaterial unit cell was also presented to enhance gain and aperture efficiency, and a co-polarized pattern diversity was presented. A real-world prototype deployment study was also carried out to demonstrate the idea. Hence the proposed modules could be potential candidates for future 5G base stations.

References

1. Rappaport TS, Sun S, Mayzus R, et al. Millimeter wave mobile communications for 5G cellular: It Will Work!, *IEEE Access*. 2013; 1: 335–349.
2. Roh W, Seol J-Y, Park J, et al. Millimeter-wave beamforming as an enabling technology for 5G cellular communications: Theoretical feasibility and prototype results. *IEEE Commun Mag*. 2014; 52(2): 106–113.
3. Pi Z, Khan F. An introduction to millimeter-wave mobile broadband systems. *IEEE Commun Mag*. 2011; 49(6): 101–107.
4. Hong W, Baek K, Lee Y, Kim Y, Ko S. Study and prototyping of practically large-scale mmWave antenna systems for 5G cellular devices. *IEEE Commun Mag*. 2014; 52(9): 63–69.
5. Cui Y, Li R, Wang P. Novel dual-broadband planar antenna and its Array for 2G/3G/LTE Base stations. *IEEE Trans Antennas Propag*. 2013; 61(3): 1132–1139.
6. Zhao Y. Dual-wideband microstrip antenna for LTE indoor base stations. *Electron Lett*. 52(8): 576–578.
7. Choubey PN, Hong W, Hao Z, Chen P, Duong T, Mei J. A wideband dual-mode SIW cavity-backed triangular-complimentary-split-ring-slot (TCSRS) antenna. *IEEE Trans Antennas Propag*. 2016; 64(6): 2541–2545.

8. Yang B, Yu Z, Dong Y, Zhou J, Hong W. Compact tapered slot antenna array for 5G millimeter-wave massive MIMO systems. *IEEE Trans Antennas Propag.* 2017; 65(12): 6721–6727.
9. Ta SX, Choo H, Park I. Broadband printed-dipole antenna and its arrays for 5G applications. *IEEE Antennas Wirel Propag Lett.* 2017; 16: 2183–2186.
10. Reddy GS, Kamma A, Kharche S, Mukherjee J, Mishra SK. Cross-configured directional UWB antennas for multidirectional pattern diversity characteristics. *IEEE Trans Antennas Propag.* 2015; 63(2): 853–858.
11. Dong Y, Choi J, Itoh T. Vivaldi antenna with pattern diversity for 0.7 to 2.7 GHz cellular band applications. *IEEE Antennas Wirel Propag Lett.* 2018; 17(2): 247–250.
12. Monavar FM, Shamsinejad S, Mirzavand R, Melzer J, Mousavi P. Beam-steering SIW leaky-wave subarray with flat-topped footprint for 5G applications. *IEEE Trans Antennas Propag.* 2017; 65(3): 1108–1120.
13. Zhu S, Liu H, Chen Z, Wen P. A compact gain-enhanced Vivaldi antenna array with suppressed mutual coupling for 5G mmWave application. *IEEE Antennas Wirel Propag Lett.* 2018; 17(5): 776–779.
14. Gan Z, Tu Z, Xie Z. Pattern-reconfigurable unidirectional dipole antenna array fed by SIW coupler for millimeter wave application. *IEEE Access.* 2018; 6: 22401–22407.
15. Friis HT. A note on a simple transmission formula. *Proc IRE.* 1946; 34(5): 254–256.
16. Dastranj A, Abiri H, Mallahzadeh A. Design of a broadband cosecant squared pattern reflector antenna using IWO algorithm. *IEEE Trans Antennas Propag.* 2013; 61(7): 3895–3900.
17. Smith DR, Schultz S, Markos P, Soukoulis CM. Determination of effective permittivity and permeability of metamaterials from reflection and transmission coefficients. *Phys Rev B.* 2002; 65: 195104.
18. Kota K, Shafai L. Gain and radiation pattern enhancement of balanced antipodal Vivaldi antenna. *Electron Lett.* 2011; 47(5): 303–304.
19. Nassar IT, Weller TM. A novel method for improving antipodal Vivaldi antenna performance. *IEEE Trans Antennas Propag.* 2015; 63(7): 3321–3324.
20. Chen L, Lei Z, Yang R, Fan J, Shi X. A broadband artificial material for gain enhancement of antipodal tapered slot antenna. *IEEE Trans Antennas Propag.* 2015; 63(1): 395–400.
21. Zhou B, Li H, Zou X, Cui T-J. Broadband and high-gain planar Vivaldi antennas based on inhomogeneous anisotropic zero-index metamaterials. *Prog Electromagn Res.* 2011; 120: 235–247.
22. Guo J, Liao S, Xue Q, Xiao S. Planar aperture antenna with high gain and high aperture efficiency for 60-GHz applications. *IEEE Trans Antennas Propag.* 2017; 65(12): 6262–6273.
23. Jang TH, Kim HY, Song IS, Lee CJ, Lee JH, Park CS. A wideband aperture efficient 60-GHz series-fed E-shaped patch antenna array with copolarized parasitic patches. *IEEE Trans Antennas Propag.* 2016; 64(12): 5518–5521.
24. Li H, Wang G, Cai T, Liang J, Gao X. Phase- and amplitude-control metasurfaces for antenna main-lobe and sidelobe manipulations. *IEEE Trans Antennas Propag.* 2018; 66(10): 5121–5129.
25. Dadgarpour A, Zarghooni B, Virdee BS, Denidni TA. One- and two-dimensional beam-switching antenna for millimeter-wave MIMO applications. *IEEE Trans Antennas Propag.* 2016; 64(2): 564–573.
26. Briqech Z, Sebak A, Denidni TA. Wide-scan MSC-AFTSA array-fed grooved spherical lens antenna for millimeter-wave MIMO applications. *IEEE Trans Antennas Propag.* 2016; 64(7): 2971–2980.
27. Dadgarpour A, Zarghooni B, Virdee BS, Denidni TA. Beam-deflection using gradient refractive-index media for 60-GHz end-fire antenna. *IEEE Trans Antennas Propag.* 2015; 63(8): 3768–3774.
28. Wani Z, Abegaonkar MP, Koul SK. Millimeter-wave antenna with wide-scan angle radiation characteristics for MIMO applications. *Int J RF Microw Comput Aided Eng.* 2019; 29(5): e21564.
29. Karthikeya, G. S., Mahesh P. Abegaonkar, and Shiban K. Koul. "Pattern diversity of path loss compensated antennas for 5G base stations." *International Journal of RF and Microwave Computer-Aided Engineering* 2019; 29(8): e21800.

6
Shared Aperture Antenna with Pattern Diversity for Base Stations

6.1 Introduction

The tremendous recent increase in data-hungry applications, especially with respect to smartphone users, has encouraged researchers in academia and leading research organizations to look for the design of hardware ecosystem at 28 GHz and beyond for future 5G cellular systems, primarily because of spectral congestion in the sub-6 GHz bands. Path loss for mmWave communication links is high; for example, for a 10 m link at 28 GHz, the path loss is 82 dB as against 60 dB for the same distance for the existing commercial 4G cellular link. The penetration losses are also high (>20 dB) for the 28 GHz for common building materials such as concrete and brick [1].

In order to establish a reasonable communication link, high-gain antennas must be deployed at the base stations and mobile terminals to compensate path loss, as noted in [2] for the proposed 5G band. But high-gain antennas would necessarily mean low beamwidth, thus leading to poor coverage. Base station antennas must have reasonably high gain, therefore high gain antennas with the smallest physical footprint would be ideal candidates for 5G base station operation in the 28 GHz band. High gain antennas could be designed using phased array schemes such as given in [3,4], but the configuration of phase shifters and the controllers for beam-locking for the communication link would be challenging to design. Also, an N-port phased array system (N>4), which would create a resultant beam at 0° and 90°, would suffer from scanning loss and hence lead to gain deterioration away from the boresight. As the beam is scanned away from 0°, the side-lobe level also increases, therefore antennas with orthogonal pattern diversity are preferred. It must be noted that pattern diversity architecture also increases the data throughput of the communication system. Shared aperture antennas could be redesigned for mmWave 5G base station to achieve high gain with pattern diversity and the smallest physical footprint. The pattern diversity topology suggested in [5] has four orthogonal ports with low mutual coupling in the 2–20 GHz band; shared aperture is not a feature. Shared aperture proposed in [6] does not have independently controllable patterns. It must also be noted that the gain variation between the two modes of operation is more than 2 dB across the bandwidth because of the dual beamforming. Gain enhancement architecture with zero index metamaterial (ZIM) loading proposed in [7] would operate only for the designed incident polarization. Similarly, the metamaterial superstrate presented in [8] is sensitive to the incident polarization, and the decoupling strategy might not be operational for orthogonal incident waves. Miniaturization techniques such as the integration of a modified circular slot ring resonator (CSRR) with the antenna would yield a compact design with a compromise in gain [9], hence proving unsuitable for base station applications. Shared aperture design targeting satellite applications [10] has a compact design but the gain variation across the ports or bands is noticeable. Thus, in order to achieve uniform gain for orthogonal ports with minimal physical footprint, a shared aperture antenna with dual-polarized zero-index metamaterial (DPZIM) unit cells is proposed. A shared aperture antenna with orthogonal pattern diversity is presented in this chapter. A dual-polarized metamaterial unit cells is designed and characterized with identical characteristics for both of the incident polarizations. The proposed unit cells are strategically integrated to the shared aperture antenna to achieve gain enhancement, gain equalization across the ports and the band, along with reduction in mutual coupling.

6.2 Shared Aperture Antenna

The deployment scenario of the 5G base station module is evident from Figure 6.1(a). In order to achieve this, a shared aperture antenna with orthogonal pattern diversity is proposed. The schematic of the proposed antenna with shared aperture and pattern diversity is depicted in Figure 6.1(b), and the corresponding photograph is shown in Figure 6.1(c). The structure is designed on Nelco NY9220 with a dielectric constant of 2.2 and 20-mil thickness. A low dielectric constant was chosen to reduce additional surface wave modes, and an electrically thin substrate would reduce cross-polarization radiation in end-fire. Typically, substrates are characterized for the dielectric constant at 10 GHz, and the manufacturer would specify the tolerance for the substrate. In this case, the dielectric constant varies within 2.2 ± 0.02. It would be good design practice to measure the dielectric constant of the available substrate in the desired frequency by making a ring resonator or a standard inset-fed patch antenna. The loss tangent could also be discovered by measuring the transmission loss for a standard 50 Ω line for a 50 mm line of the substrate in a properly calibrated vector network analyser. It is also recommended to measure the thickness of the substrate using digital Vernier calipers or a digital screw gauge. The thickness of copper would be 15–35 μm. The thickness of the copper would not affect the performance of the antenna, but would decide its power handling capacity.

The feed line is a standard 50 Ω line of width 1.5 mm, in series with a quarter-wave transformer (QWT) of 96 Ω with a line width of 0.5 mm transitioning to the microstrip with a slotline transition of 194 Ω. Two antennas with identical flare angles of the radiating aperture are merged into a single shared aperture with a port-to-port distance of 30 mm. The elements could be designed to be electrically closer, but this would result in the degradation of mutual coupling and radiation characteristics, hence the chosen topology is a compromise between mutual coupling, beam integrity across band and gain with least physical footprint, for a compact antenna with pattern diversity across the 27–30 GHz band for 5G base stations.

The flare angle of the radiating aperture is designed for optimal beamwidth in the end-fire. The shared aperture increases the mutual coupling between the two orthogonal ports. The structural asymmetry leads to variation in radiation pattern variation across the band. The shared aperture is designed with tapered slot antennas, since the beamwidth could be engineered for the application in hand. Tapered slot antenna (TSA) elements also offer high beam integrity throughout the operational spectrum. Previously reported orthogonal pattern diversity antenna modules have gain variation across the band and across the ports

Shared Aperture Antenna with Pattern Diversity for Base Stations   151

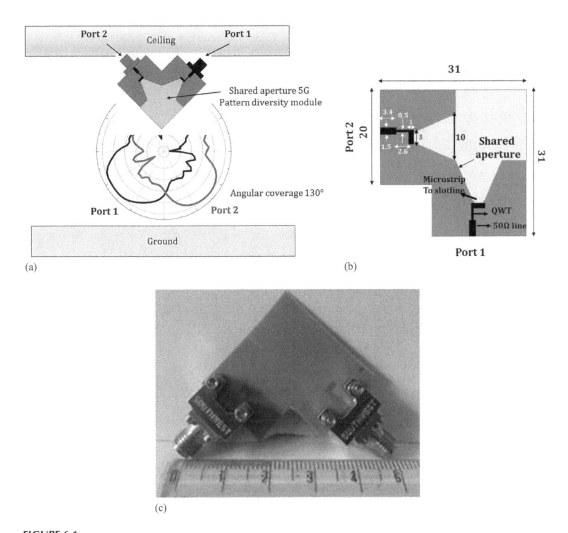

FIGURE 6.1
(a) Typical deployment scenario of the antenna, (b) Schematic of the proposed shared aperture antenna and (c) Photograph of the shared aperture antenna [24].

because of the utilization of different class of antennas for orthogonal beams. Leaky wave topology would be unable to maintain radiation pattern integrity when the beam is excited for a 90° tilt with respect to boresight. Similarly, phased array architecture reported in the literature suffers from a scanning loss of almost 2 dB for a beam tilt of 45°, hence for a beam scan at 90° the scanning loss would be higher, leading to increased degradation of the patterns.

The E-field plots for both the ports at 28 GHz are illustrated in Figure 6.2. The quasi-cylindrical wavefront at the edge of the physical aperture is evident, which creates a larger error in the phase front, leading to a lower gain for both the ports. The mutual coupling is also evident from the figures.

The simulated and measured S-parameters for both of the ports are depicted in Figure 6.3. The impedance bandwidth for the shared aperture element is 27–30 GHz (10%). The impedance bandwidth could be enhanced by replacing the stepped impedance

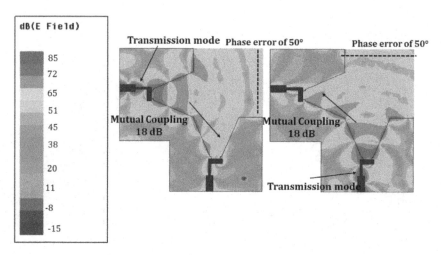

FIGURE 6.2
E-fields for port excitation at 28 GHz, (a) port 2 and (b) port 1 [24].

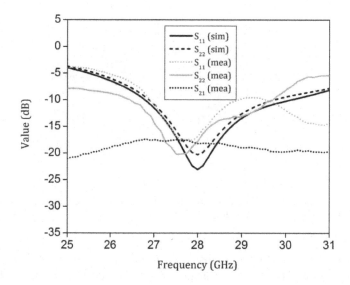

FIGURE 6.3
S-parameters of the proposed shared aperture antenna [24].

transformer with a wideband balun, but this would lead to pattern distortion at the higher end of the spectrum, leading to significant variation in gain. The impedance characteristics are similar for both of the orthogonal ports. The measured mutual coupling is in the range −17 dB to −19 dB in the 27–30 GHz frequency band. The mutual coupling is high because of the shared aperture and the ports being electrically close. The electrical distance is designed for an optimal mutual coupling and beam integrity.

6.3 DPZIM Design and Characterization

The end-fire gains of the proposed shared aperture antenna vary between 7 and 8 dBi. In order to enhance gain with the same physical footprint, the integration of metamaterial unit cells is investigated. Typical gain enhancement technique such as quasi-Yagi topology is strongly polarization sensitive and hence, in the context of shared aperture, gain enhancement would be achieved in one of the orientations in addition to degrading the far field patterns of the orthogonal port. The zero-index metamaterial unit cells proposed in the literature are also strongly polarization sensitive, therefore the integration of these unit cells might not be optimal solution for simultaneous gain enhancement across the ports with a reduction in mutual coupling. The design evolution of the unit cell is depicted in Figure 6.4.

To achieve gain enhancement in dual polarization modes, the insertion loss ($|S_{21}|$) must be minimal and the unit cell must create a reasonable alteration of the phase of the incoming wave. As observed in the S-parameter curves, the circular ring has a high insertion loss. Design B has a relatively lower insertion loss, but the relative phase difference is insignificant, hence the proposed design has the least insertion loss with a substantial phase difference, as seen in the figure. The orthogonally symmetric slots act as capacitors, and the arc-shaped stubs act as inductors, leading to a near zero refractive index, as illustrated in the extracted parameters in Figure 6.5 in the frequency of interest (27–30 GHz).

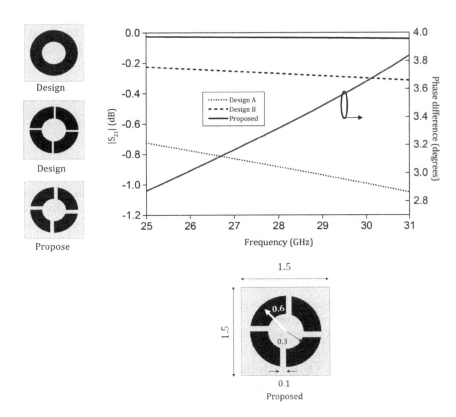

FIGURE 6.4
Design evolution and the proposed unit cell [24].

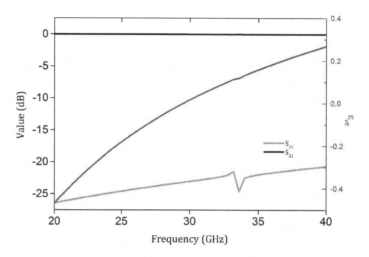

FIGURE 6.5
Extracted parameters of the unit cell [24].

The parameters are extracted by the method described in [15]. The orthogonally symmetric slots aid in the polarization insensitive action of gain enhancement. The capacitive slots could be radially increased for a multiport polarization, but the gain yield post integration would be minimal. The gain enhancement would be optimal when the polarization of the incident E-field matched the slots of the unit cell, which in turn would reduce the phase error of the E-plane across the physical aperture and hence effectively increase the radiating aperture.

6.4 Shared Aperture Antenna with DPZIM

The E-field plots post integration with the DPZIM unit cells is shown in Figure 6.6. The planarization of the cylindrical wavefront is evident for both the ports compared to the plain shared aperture antenna presented in section 6.2. The strategic loading of the unit cells is designed for gain enhancement, and gain equalization of the ports with simultaneous reduction in mutual coupling between the ports.

The polarization insensitive nature is operational in the shared aperture topology. The orthogonal slots of the unit cell match the orthogonal polarization of the ports of the shared aperture antenna. The normalized phase distribution with and without the integration of DPZIM is demonstrated in Figure 6.7. The phase error across 1λ has been reduced from 50° to 20°, hence leading to gain enhancement of more than 2 dB across the band and the ports. The schematic of the proposed element, along with the photograph, is shown in Figure 6.8.

$|S_{11}|$ and $|S_{22}|$ of the respective ports remain almost unaltered post integration with the DPZIM, as illustrated in Figure 6.9 primarily because of the minimal effect on the microstrip to slotline transition of the antenna. The measured mutual coupling for the ZIM loaded shared aperture antenna is in the range −23 dB to −20 dB in the 27–30 GHz frequency band, hence a decrease of more than 3 dB across the band is observed. Since the mutual coupling is reduced, this would lead to higher isolation between the ports, leading

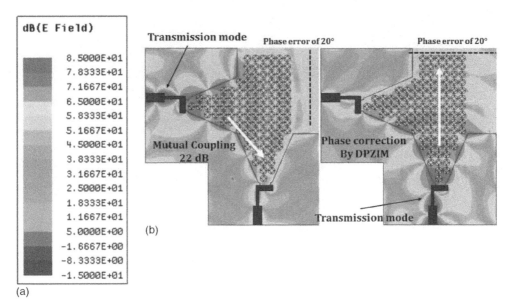

FIGURE 6.6
E-field plots at 28 GHz, (a) port 1 and (b) port 2 [24].

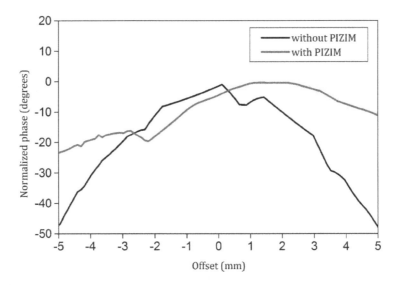

FIGURE 6.7
Normalized phase distribution at 28 GHz [24].

to radiation patterns with higher beam integrity in the frequency of operation. The simulated and measured end-fire radiation patterns, with pattern diversity at 28 and 30 GHz, are shown in Figure 6.10(a) and 6.10(b). It must be observed that beamwidth is 40° ± 5° for each of the ports in the frequency of operation. The patterns are stable across the band for both of the ports. The front-to-back ratio is more than 15 dB and the patterns are orthogonal, hence proving its utility in a typical 5G base station.

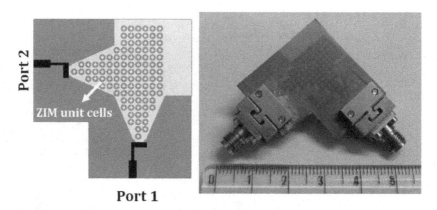

FIGURE 6.8
Schematic and photograph of the proposed DPZIM antenna [24].

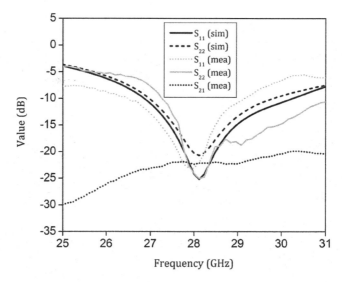

FIGURE 6.9
S-parameters of the proposed DPZIM antenna [24].

The discrepancy between the simulated and measured radiation characteristics is primarily a result of the effect of lossy adapters and other transitions utilized for measurements. The flaring angle of the shared aperture decides the beamwidth. The current topology is optimized for low beamwidth, consequently leading to higher gain, which is essential to maintain a fair communication link budget in a standard communication setup to compensate high path loss in the 27–30 GHz frequency band.

The simulated and measured gains for each port are illustrated in Figure 6.11. It is observed that the end-fire gains of the orthogonal ports is 7–8 dBi without the integration of ZIM unit cells, and the gain variation is almost 1 dB between the ports because of the asymmetry in the dielectric loading of the shared aperture. The simulated gain after integrating with ZIM is in the range of 9.2–9.6 dBi for both the ports for the 27–30 GHz

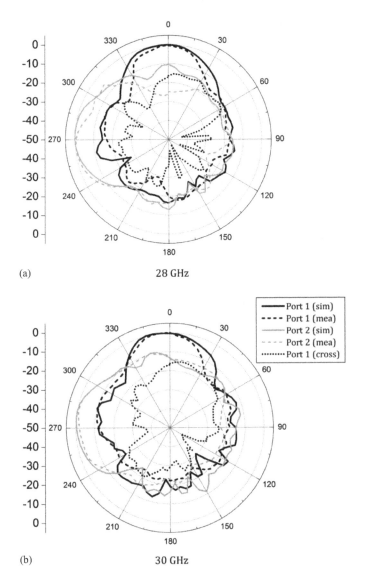

FIGURE 6.10
Radiation patterns at (a) 28 GHz and (b) 30 GHz [24].

frequency range, and measured gain is in the range 8.5–9.2 dBi for the 27–30 GHz frequency band. The discrepancy between simulated and measured gain could be attributed to frequency sensitive adapters and other transitions utilized in the standard gain transfer method.

Table 6.1 illustrates the features of the proposed pattern diversity module against other reported designs. It is evident that the 1 dB gain bandwidth is 35%, indicating a wideband high pattern integrity behaviour for orthogonal pattern diversity with minimal effective radiating volume. The proposed metamaterial unit cells operate with orthogonal incident polarizations, compared to most of the reported articles, which have strong polarization sensitivity. Leaky wave topology would be unable to maintain radiation pattern integrity when the beam is excited for a 90° tilt with respect to boresight [12]. Conventional gain

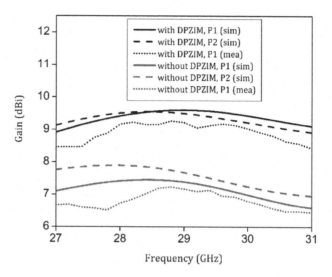

FIGURE 6.11
End-fire gains of the proposed antennas [24].

TABLE 6.1

Comparison with Reported Designs

REF	F	G	GBW	MC	ERV	PD
[3]	28	9	24.6	NA	0.138	+40°, −40°
[4]	28	6	24.2	20	0.01	0°, 45°
[5]	16	8.5	22.2	25	0.14	0°, 180°, ±90°
[7]	10	14	23.2	NA	0.04	No
[8]	2.5	6.1	NA	42	0.116	No
[11]	5.8	4	2.6	20	0.016	0°, ±90°
[16]	60	12	11.6	20	1.09	+35°, −35°
[17]	60	20	21.8	30	68.64	0°,+30°,−30°
[18]	64	11	6.1	NA	0.08	No
[19]	28	11	10.9	16	0.05	0°, +30°, −30°
Proposed	28	9	35	22	0.09	0°, 90°

Note F = Centre Frequency (GHz), G = Gain (dBi), GBW = 1 dB Gain Bandwidth (%), MC = Mutual Coupling (dB), ERV = Effective radiating volume (λ_0^3), PD = Pattern diversity.

enhancement techniques such as loading the antenna with parasitic stubs as in typical Yagi-type antennas [13,14] the parasitic is strongly polarization sensitive and hence would fail to yield

gain enhancement in an orthogonal pattern diversity scenario, as illustrated in Figures 6.12(a) and 6.12(b). It must also be noted that the metamaterial unit cells suggested in the following references [7,8,18,19] operate when the polarization of the incident wave coincides with the polarization of the structure. The proposed dual polarized ZIM unit cells operate for both polarizations of the incident wave, as depicted in Figure 6.12(c). Multiple antenna systems proposed in [20–23] lack orthogonal beam switching.

Shared Aperture Antenna with Pattern Diversity for Base Stations 159

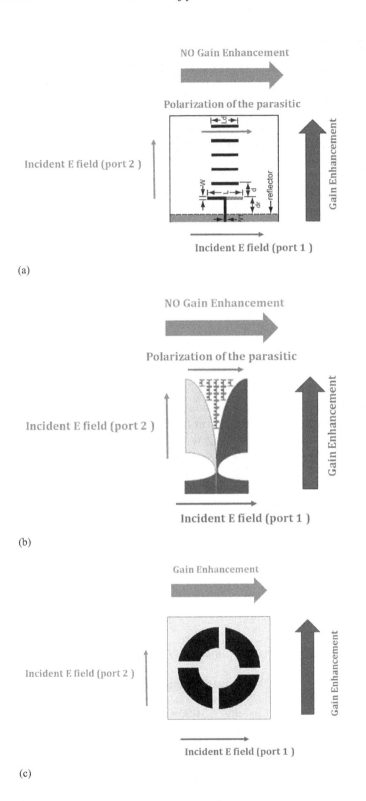

FIGURE 6.12
(a) Schematic for orthogonal pattern diversity in a Yagi scenario [13], (b) Schematic for orthogonal pattern diversity in a ZIM scenario [18] and (c) Schematic for orthogonal pattern diversity in the proposed DPZIM context.

6.5 Design Guidelines for High-Gain Dual-Polarized Antenna Module

1. A compact dual-polarized antenna module must be designed, which could be orthogonal wire antennas or aperture antennas with individual feeds, but the elements must be electrically close to realize the smallest physical footprint.
2. The metamaterial unit cell specifications described in section 4.4 are also valid for a dual-polarized design. In addition to these characteristics, the unit cell must offer them for both the orthogonal incident polarizations to achieve simultaneous phase correction for the corresponding port excitation.
3. The strategic integration of dual-polarized metamaterial unit cells would yield gain enhancement for both the ports with a consequent decrease in the mutual coupling between the ports. The pattern integrity is also decided by the characteristics of the unit cells across the band.

6.6 Conclusion

A shared aperture antenna with orthogonal polarization diversity was presented in this chapter. A dual-polarized metamaterial unit cell was proposed and characterized for gain enhancement. The metamaterial loading was designed for wideband high-pattern integrity with orthogonal pattern diversity.

References

1. T. S. Rappaport et al., "Millimeter wave mobile communications for 5G cellular: It will work!," *IEEE Access*, 1, 335–349, 2013.
2. W. Hong, K. Baek, Y. Lee, Y. Kim, and S. Ko, "Study and prototyping of practically large-scale mmWave antenna systems for 5G cellular devices," *IEEE Communication Magazine*, 52(9), 63–69, September 2014.
3. B. Yang, Z. Yu, Y. Dong, J. Zhou, and W. Hong, "Compact tapered slot antenna array for 5G millimeter-wave massive MIMO systems," *IEEE Transactions on Antennas Propagation*, 65(12), 6721–6727, December 2017.
4. S. X. Ta, H. Choo, and I. Park, "Broadband printed-dipole antenna and its arrays for 5G applications," *IEEE Antennas Wireless Propagation Letters*, 16, 2183–2186, 2017.
5. G. S. Reddy, A. Kamma, S. Kharche, J. Mukherjee, and S. K. Mishra, Cross-configured directional UWB antennas for multidirectional pattern diversity characteristics," *IEEE Transactions on Antennas Propagation*, 63(2), 853–858, February 2015.
6. Y. Dong, J. Choi, and T. Itoh, "Vivaldi antenna with pattern diversity for 0.7 to 2.7 GHz cellular band applications," *IEEE Antennas Wireless Propagation Letters*, 17(2), 247–250, February 2018.
7. B. Zhou, H. Li, X. Zou, and T.-J. Cui, "Broadband and high-gain planar Vivaldi antennas based on inhomogeneous anisotropic zero-index metamaterials," *Progress in Electromagnetic Research*, 120, 235–247, 2011.

8. F. Liu, J. Guo, L. Zhao, X. Shen, and Y. Yin, "A meta-surface decoupling method for two linear polarized antenna array in sub-6 GHz base station applications," *IEEE Access*, 7, 2759–2768, 2019.
9. A. K. Singh, M. P. Abegaonkar, and S. K. Koul, "Miniaturized multiband microstrip patch antenna using metamaterial loading for wireless application," *Progress in Electromagnetic Research C*, 83, 71–82, 2018.
10. C. Mao, S. Gao, Y. Wang, Q. Luo, and Q. Chu, "A shared-aperture dual-band dual-polarized filtering-antenna-array with improved frequency response," *IEEE Transaction on Antennas Propagation*, 65(4), 1836–1844, April 2017.
11. Y. Sharma, D. Sarkar, K. Saurav, and K. V. Srivastava, "Three-element MIMO antenna system with pattern and polarization diversity for WLAN applications," *IEEE Antennas Wireless Propagation Letters*, 16, 1163–1166, 2017.
12. F. M. Monavar, S. Shamsinejad, R. Mirzavand, J. Melzer, and P. Mousavi, "Beam-steering SIW leaky-wave subarray with flat-topped footprint for 5G applications," *IEEE Transactions on Antennas Propagation*, 65(3), 1108–1120, March 2017.
13. R. A. Alhalabi and G. M. Rebeiz, "High-gain Yagi-Uda antennas for millimeter-wave switched-beam systems," *IEEE Transactions on Antennas Propagation*, 57(11), 3672–3676, November 2009.
14. R. A. Alhalabi and G. M. Rebeiz, "Differentially-fed millimeter-wave Yagi-Uda antennas with folded dipole feed," *IEEE Transactions on Antennas Propagation*, 58(3), 966–969, March 2010.
15. D. R. Smith, S. Schultz, P. Markos, and C. M. Soukoulis, "Determination of effective permittivity and permeability of metamaterials from reflection and transmission coefficients," *Physical Review B*, 65, 1951041–1951045, 2002.
16. A. Dadgarpour, B. Zarghooni, B. S. Virdee, and T. A. Denidni, "One- and two-dimensional beam-switching antenna for millimeter-wave MIMO applications," *IEEE Transactions on Antennas Propagation*, 64(2), 564–573, February 2016.
17. Z. Briqech, A. Sebak, and T. A. Denidni, "Wide-scan MSC-AFTSA array-fed grooved spherical lens antenna for msillimeter-wave MIMO applications," *IEEE Transactions on Antennas Propagation*, 64(7), 2971–2980, July 2016.
18. M. Sun, Z. N. Chen, and X. Qing, "Gain enhancement of 60-GHz antipodal tapered slot antenna using zero-index metamaterial," *IEEE Transactions on Antennas Propagation*, 61(4), 1741–1746, April 2013.
19. Z. Wani, M. P. Abegaonkar, and S. K Koul, "Millimeter-wave antenna with wide-scan angle radiation characteristics for MIMO applications," *International Journal of RF and Microwave Computer Aided Engineering*, 29, e21564, 2019.
20. A. Iqbal et al., "Electromagnetic bandgap backed millimeter-wave MIMO antenna for wearable applications," *IEEE Access*, 7, 111135–111144, 2019.
21. M. Al-Hasan, I. Ben Mabrouk, E. R. F. Almajali, M. Nedil, and T. A. Denidni, "Hybrid isolator for mutual-coupling reduction in millimeter-wave MIMO antenna systems," *IEEE Access*, 7, 58466–58474, 2019.
22. Y. Li, L. Ge, M. Chen, Z. Zhang, Z. Li, and J. Wang, "Multibeam 3-D-printed luneburg lens fed by magnetoelectric dipole antennas for millimeter-wave MIMO applications," *IEEE Transactions on Antennas and Propagation*, 67(5), 2923–2933, May 2019.
23. Y. Li and K.-M. Luk, "60-GHz dual-polarized two-dimensional switch-beam wideband antenna array of aperture-coupled magneto-electric dipoles," *IEEE Transactions on Antennas Propagation*, 64(2), 554–563, February 2016.
24. K. Gulur Sadananda, M. P. Abegaonkar, and S. K. Koul, "Gain equalized shared-aperture antenna using dual-polarized ZIM for mmWave 5G base stations," *IEEE Antennas and Wireless Propagation Letters*, 18(6), 1100–1104, June 2019.

7
Co-Design of 4G LTE and mmWave 5G Antennas for Mobile Terminals

7.1 Introduction

The increasing demand for data rates has provoked researchers across the world to design cellular hardware ecosystems for backward compatibility. The evolution in smartphones has provoked microwave engineers to develop sub-systems that are compatible with existing protocols. New technologies have evolved and are producing portable devices that will include Long Term Evolution (LTE) for both voice and data applications.

LTE technology typically consists of three operating low-frequency bands: LTE700 (698–787 MHz), LTE2300/Class 40 band (2300–2400 MHz) and LTE2500/Class 7 band (2500–2690 MHz) [1]. Future transceivers must also accommodate mmWave 5G hardware. Frequencies for mmWave 5G are expected to be around 28 GHz for the design of 5G antennas [2]. Achieving orthogonal pattern diversity with a low physical footprint is challenging. 4G LTE and mmWave 5G MIMO antennas on the same module presents an important problem when addressing the backward compatibility of future smartphones [3].

The designs reported in [2,4] have demonstrated co-design of 4G LTE and mmWave 5G antennas, but their effective post-integration radiating volume is electrically large. Conformal antennas have been designed extensively for various applications where the surface is not flat (for example, the singly curved antenna reported in [5]). Compact design is necessary because of a lack of real estate. In this chapter, the co-design of 4G LTE and mmWave 5G antennas is presented.

7.2 Miniaturization Techniques for Antenna Size Reduction

The half wavelength at 28 GHz is 5 mm, which means that antennas with a physical footprint able to fit inside a smartphone panel could be designed without compromise in gain or radiation efficiency. On the other hand, 4G LTE operates in the 0.7–2.7 GHz band. Half-wavelengths corresponding to this band are in the range 214–255 mm. This indicates that conventional half-wavelength printed dipoles or patch antennas would not be suitable for integration with the mobile terminal. In the future, 4G antennas must be miniaturized for a compromise in the gain.

Some popular miniaturization techniques are as follows:

- Shorting posts could be integrated in the substrate of the printed dipole to increase the effective length of the antenna; miniaturization of up to 50% could be achieved, with a compromise in radiation efficiency.
- Meandering of the primary radiator. The dipole arms are conventionally flat; if these could be meandered then the effective radiating length would be increased for a given physical footprint. Bandwidth is severely limited in this strategy.
- High dielectric constant substrates could be used when designing the antenna. Miniaturization of up to 40% could be achieved, depending on the magnitude of the dielectric constant. These embedded antennas suffer from relatively higher loss, leading to poor radiation efficiency.
- Parasitic element loading is another common technique used to increase the effective length. Multi-band operation could also be obtained in this technique by modifying the geometry of the parasitic.
- Matching circuits could be integrated with the antenna element for miniaturization. This a popular technique in the industry. The input impedance of an antenna that fits into the mobile panel would be used to co-design the antenna with a tunable LC circuit [6].

7.3 Conformal 4G LTE MIMO Antenna Design

The 4G LTE antenna simulations are carried out in computer simulation technology (CST) Microwave Studio. The 4G MIMO antenna system is designed on the Rogers 5870 substrate with dielectric constant of 2.33 and loss tangent of 0.0012. To minimize surface wave modes, a substrate of low dielectric constant is chosen. The substrate has a thickness of 10 mil, with which it is easier to achieve the conformity. Flexible substrates such as polyethylene terephthalate (PET) and polycarbonate have a high loss tangent, which will result in additional gain deterioration [7]; 5 mil substrates are more flexible than the proposed substrate for conformal designs but require additional scaffolding [8].

7.3.1 CRLH-Based Conformal 4G LTE Antenna

Several techniques have been investigated in [9–11] to miniaturize the antenna, but all decrease the antenna's efficiency. A general composite right/left handed (CRLH)-based unit cell comprises series capacitance (C_L), series inductance (L_R), shunt inductance (L_L) and shunt capacitance (C_R). Resonant frequencies for series and shunt configuration are given by [12]:

$$\omega_{se} = \frac{1}{\sqrt{L_R C_L}} \text{ rad/s} \qquad (7.1)$$

$$\omega_{sh} = \frac{1}{\sqrt{L_L C_R}} \text{ rad/s} \qquad (7.2)$$

The metamaterial (MTM) inspired unit cell resonates at a specific frequency, which satisfies the following resonance condition as per the theory of open-ended resonator with the CRLH transmission line [13]: $\beta_n = n\pi/p$, where β_n is the phase constant, n is the mode of resonance and p is the length of the period. When $n = 0$, the zeroth order resonating (ZOR) mode can be attained in which the physical size of resonator is independent of the frequency of operation [14] as against conventional strongly resonant antennas which have $\lambda/2$ electrical size. Thus the miniaturization of the antenna can be achieved in the ZOR mode. The proposed 4G LTE monopole antenna deploying CRLH MTM is depicted in Figure 7.1.

The planar CRLH 4G LTE antenna has the physical size of $20 \times 20 \times 0.254$ mm^3, which is then bent by 90° to decrease the footprint of the antenna and hence can easily be integrated on a typical smartphone. The proposed conformal 4G antenna has minimal effective radiating volume (ERV) after being bent, which is $0.00004\ \lambda_0^3$. The proposed prototype is fabricated and respective photographs are shown in Figures 7.2(a) and 7.2(b), front and back views, respectively.

The equivalent circuit model for the proposed CRLH-based 4G LTE antenna is depicted in Figure 7.3, in which series resistance (R) and shunt conductance (G) are also presented. The ZOR frequency depends only on the shunt parameters [14]. Thus the proposed antenna takes mainly into consideration the shunt parameters L_L and C_R. Also, the fractional bandwidth (BW) of the open-ended ZOR antenna is calculated by [12]:

$$BW = G \times \sqrt{\frac{L_L}{C_R}} \qquad (7.3)$$

To achieve effective miniaturization without a decrease in the fractional bandwidth of this ZOR 4G LTE antenna, only L_L is increased rather than C_R, which would result in the reduction of bandwidth. Since meander lines are connected in parallel, asymmetric insertion is incorporated in comparison with their symmetric counterpart, which results in an

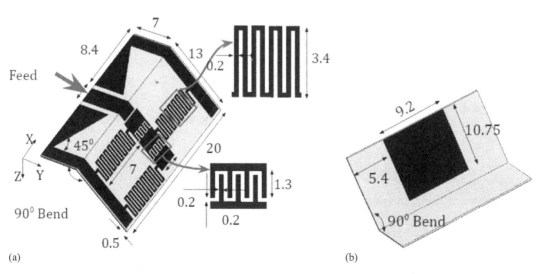

(a)   (b)

FIGURE 7.1
Schematics of the proposed 4G LTE antenna (a) Top plane and (b) Back plane (dimensions in mm) [35].

FIGURE 7.2
Photographs of the proposed prototype (a) Front view and (b) Back view [35].

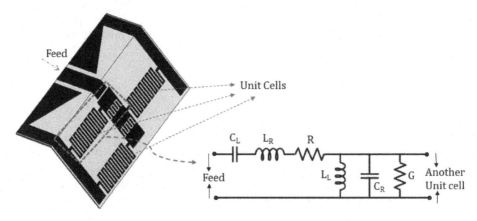

FIGURE 7.3
Equivalent circuit model for the CRLH unit cell of the proposed 4G LTE antenna [35].

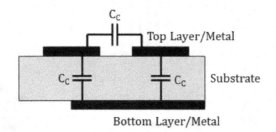

FIGURE 7.4
Schematics demonstrating the role of the bottom patch in impedance matching [35].

increase of L_L. A localized ground behind the radiator is introduced to enhance impedance matching. Coupling capacitance (C_C) is increased by placing a patch in the bottom plane, as illustrated in Figure 7.4. Thu, impedance matching can be achieved by adjusting the position and size of the bottom patch. Optimization of the equivalent circuit parameters is carried out for the desired resonance.

For electrically small antennas, because of low radiation resistance and high capacitive reactance, impedance matching is challenging [15]. To mitigate this effect, coplanar waveguide (CPW) feeding is introduced, which results in a decrease in capacitive reactance by reducing the cross-section area between the trace and the ground. In addition, a high impedance transformer line is introduced, which results in an increase in radiation resistance. CPW feeding is also introduced because of the flexibility of enhancing L_L by inserting a number of meander lines. Also, because of the higher separation between the CPW ground and the radiation patches, the C_R parameter is reduced, which contradicts the microstrip feeding technique. A microstrip-fed structure is not the optimal choice to achieve miniaturization in ZOR antennas, since parameters L_L and C_R cannot be controlled in such a structure by the geometry of the metal trace. Shorting pins could be used for miniaturization but they would lead to complexity in fabrication with narrow bandwidths.

According to standard CPW calculations [16], the characteristic impedance of the feed line in the proposed antenna is 50 Ω. The CPW ground is tapered in order to increase the impedance bandwidth (BW). Also, the CPW-fed antenna will result in the least number of discontinuities after conforming by 90°, compared to the microstrip feeding, because the latter has a copper plane on the ground which will also be at maximum strain, thus causing impedance mismatch [17].

The simulated and measured input reflection coefficient ($|S_{11}|$) of the 4G LTE planar as well as of the conformal antenna is shown in Figure 7.5. The measured results were carried out using the Agilent PNA E8364C vector network analyser. The conformal 4G LTE antenna covers class 7 LTE band. The measured impedance bandwidth of the proposed conformal 4G LTE antenna is from 2.5 to 2.65 GHz, with the fractional bandwidth of the proposed antenna being about 5.8%, which is greater than the other designs as listed in Table 7.1. For the conformal miniaturized antenna, reduction of impedance bandwidth can be seen from planar to conformal geometry. This is because of the discontinuity that occurs in the antenna while making a 90° bend. The disparity can be observed in the simulated and measured data, which may be a result of manufacturing tolerances. Also, the frequency shift may be caused by the inhomogeneous dielectric constant of the substrate [13].

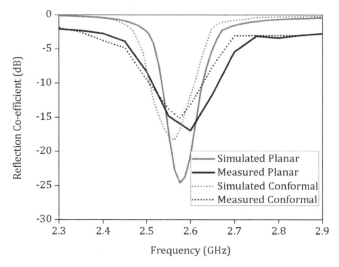

FIGURE 7.5
Input reflection co-efficient of the 4G LTE antenna [35].

TABLE 7.1
Comparison of the Proposed 4G Antenna with Reported Articles

REF	ERV	FOP	FB	FT	G	RE	Conformal
[11]	0.00023	2.88	NA	Microstrip	−0.82	46	No
[13]	0.04	5.83	NA	Microstrip	5	59.7	No
[14]	0.00005	1.22	5	Microstrip	1.73	73.4	No
[18]	0.00013	2.5	2.4	CPW	1.1	NA	No
[19]	0.00003	1.78	3	CPW	−0.15	70	No
[20]	0.00028	2.33	0.6	CPW	1.08	62	No
[21]	0.00023	3.38	<1	Microstrip	NA	NA	No
[22]	0.001	3.3	3	Coaxial	0.79	65.8	No
[23]	0.0001	2.5	2	CPW	2.56	69.2	No
[24]	NA	3.1	0.7	NA	1.3	77.4	No
PW	0.00004	2.6	5.8	CPW	2	85	Yes

Note REF = References, ERV = Effective radiating volume ($\lambda 03$), FOP = Frequency of operation (GHz), FB = Fractional bandwidth (%), FT = Feeding type, G = Gain (dBi), RE = Radiation efficiency (%), NA = Not available and PW = Proposed work.

The simulated and measured radiation patterns in both the principal planes of the proposed 4G LTE antenna are illustrated in Figure 7.6 at 2.6 GHz. The measured results were carried out in an anechoic chamber. As can be seen from the figure, the conformal antenna achieves the omnidirectional radiation pattern. Discrepancies between the simulated and measured results are because of the poor absorptivity of oblique incidence in the anechoic chamber. The measured cross-polarization radiation patterns in both the planes are 25 dB less than the co-polarization radiation pattern, indicating strongly linearly polarized antenna.

The simulated and measured realized gain of the proposed antenna varies within the operating band 1.6 to 2.2 dBi, which indicates a high gain yield for the electrical size of the proposed antenna. The simulated radiation efficiency of the proposed 4G LTE antenna has the range within 75%–90%, which is caused by the utilization of electrically thin substrate. Realized gain and radiation efficiency plots are collectively shown in Figure 7.7.

7.3.2 Compact CRLH-Based Conformal 4G LTE MIMO Antenna

The proposed 4G LTE MIMO antenna system consists of electrically close two-CRLH-based conformal antennas. The two conformal antennas, which are separated by a distance of 0.3 λ_0 can easily be placed along the edge of a typical smartphone. A photograph of the fabricated prototype with isometric view is depicted in Figure 7.8.

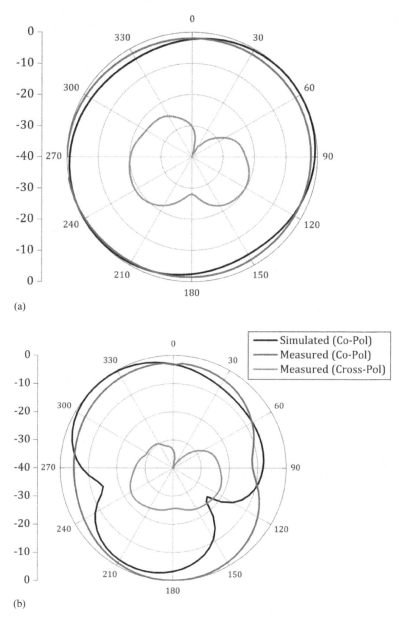

FIGURE 7.6
Simulated and measured far-field radiation patterns in (a) YZ plane and (b) XY plane at 2.6 GHz [35].

The simulated and measured isolation between two elements of the proposed 4G LTE MIMO is more than 15 dB across the whole operating band, as shown in Figure 7.9(a). Discrepancies between the simulated and measured results can be observed, which result from fabrication tolerances, substrate inhomogeneity and connector modelling. The three-dimensional (3D) realized gain of two-element proposed MIMO antenna

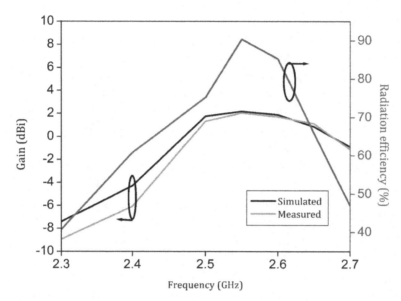

FIGURE 7.7
Peak gain and radiation efficiency of the proposed antenna [35].

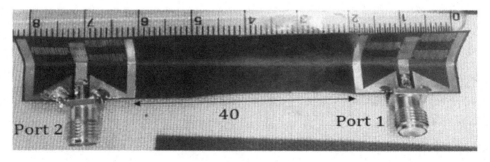

FIGURE 7.8
Photograph of the proposed 4G LTE MIMO antenna [35].

at 2.6 GHz is shown in Figure 7.9(b). The beam tilt is observed as a result of the electrical offset of antennas with respect to the phase centre.

For MIMO antennas, the Envelope Correlation Coefficient (ECC) is an important performance metric. The ECC should be minimal for the optimal performance of the antenna module. The ECC of the proposed 4G LTE MIMO antenna module is less than 0.04, as shown in Figure 7.10, which is minimal.

Co-Design of 4G LTE and mmWave 5G Antennas for Mobile Terminals 171

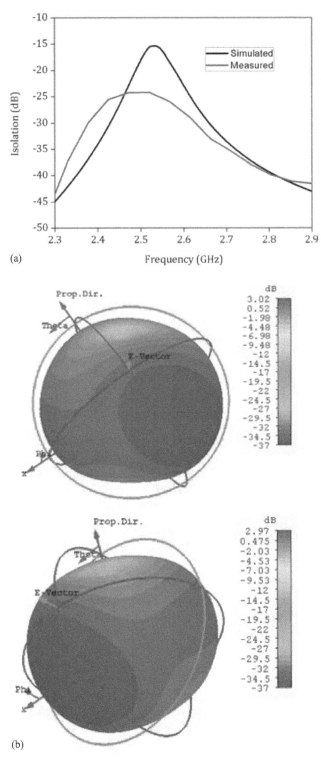

FIGURE 7.9
Isolation between the ports for 4G LTE (b) 3D radiation patterns for 4G LTE [35].

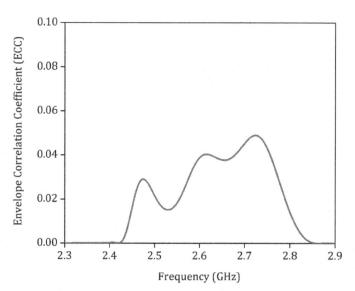

FIGURE 7.10
ECC of the proposed CRLH-based 4G LTE MIMO antenna [35].

7.4 Conformal mmWave 5G MIMO Antenna

Typical phased array designs would suffer from scanning loss when the beam is tilted away from the boresight, hence it is not feasible to design a phased array system which works at both 0° and 90°. In order to accommodate both the landscape and portrait modes, the conformal antennas are placed orthogonally at the edge of the smartphone, and beam switching could be performed when the user orients the smartphone in either landscape or portrait mode. Since the conformal antennas are placed orthogonally, the gain and the pattern integrity is maintained in either of the modes.

The proposed conformal antenna operating at 28 GHz is depicted in Figure 7.11. Its corresponding input reflection coefficient and the forward gains are depicted in Figure 7.12, followed by its radiation patterns in Figure 7.13. The antenna is a standard inset-fed microstrip patch design on Nelco NY9220 with ε_r of 2.2 and a loss tangent of 0.0009. The loss tangent plays a crucial role in the gain yield of the antenna, especially at 28 GHz. The flexible design of [7] has a loss tangent of 0.022 which reduces the gain by 2 dB. The antenna is conformed to a 90° bend, to fit into a typical mobile terminal.

Orthogonal conformal pattern diversity is also proposed, as seen in Figure 7.14. Because of the orthogonal topology of the mmWave 5G antennas in MIMO configuration, isolation of less than 40 dB is achieved in the whole operating band, as shown in Figure 7.15. Also, ECC is less than 0.00001, as depicted in Figure 7.16, which is very low. It is evident from Table 7.2 that the proposed architecture yields high gain with orthogonal pattern diversity with minimal footprint.

Co-Design of 4G LTE and mmWave 5G Antennas for Mobile Terminals 173

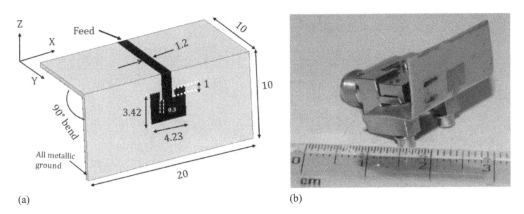

FIGURE 7.11
(a) Schematic and (b) Photograph of the proposed antenna [35].

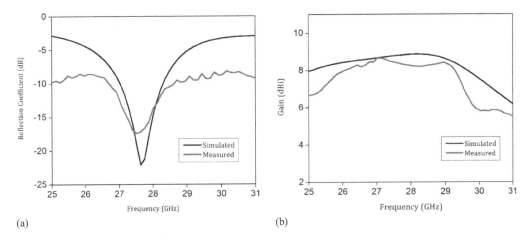

FIGURE 7.12
(a) Reflection coefficient and (b) Forward gain of the 5G antenna [35].

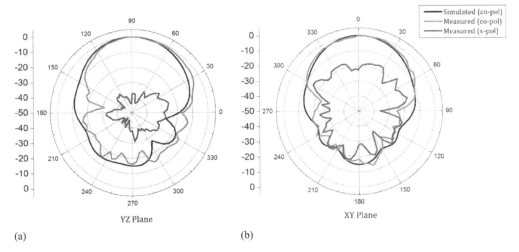

FIGURE 7.13
Patterns in (a) YZ plane and (b) XY plane at 28 GHz [35].

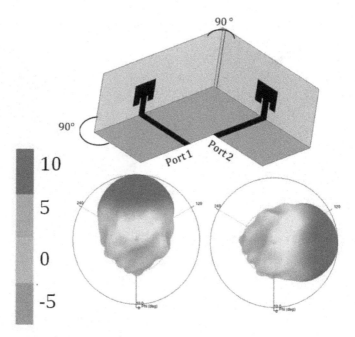

FIGURE 7.14
Orthogonal pattern diversity architecture and its 3D patterns at 28 GHz [35].

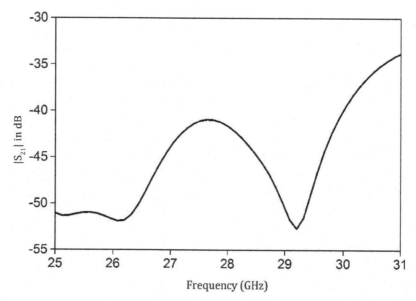

FIGURE 7.15
Isolation plot of the proposed mmWave 5G MIMO antenna [35].

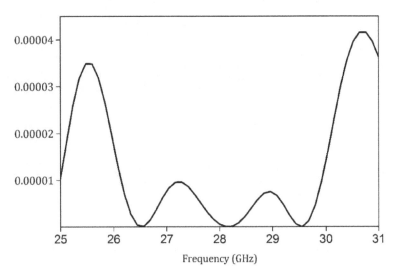

FIGURE 7.16
ECC of the proposed mmWave 5G MIMO antenna [35].

TABLE 7.2

Comparison of mmWave 5G MIMO with other Reported Articles

REF	F	G	Con.	MC	ERV	PD
[25]	28	9	No	NA	0.138	+40°, −40°
[26]	28	6	No	20	0.01	0°, 45°
[27]	16	8.5	No	25	0.14	0°, 180°, ±90°
[28]	10	14	No	NA	0.04	No
[29]	2.5	6.1	No	42	0.116	No
[30]	5.8	4	No	20	0.016	0°, ±90°
[31]	60	12	No	20	1.09	+35°, −35°
[32]	60	20	No	30	68.64	0°, +30°, −30°
[33]	64	11	No	NA	0.08	No
[34]	28	11	No	16	0.05	0°, +30°, −30°
PW	**28**	**9**	**Yes**	**35**	**0.16**	**0°, 90°**

Note REF = References, F = Centre Frequency (GHz), G = Gain (dBi), Con. = Conformal, MC = Mutual Coupling (dB), ERV = Effective radiating volume (λ_0^3), PD = Pattern diversity and PW = Proposed work.

7.5 Corner Bent Integrated Design of 4G LTE and mmWave 5G Antennas

Antenna designs and simulations are carried out in computer simulation technology (CST) Microwave Studio (MS). Proposed 4G LTE and mmWave 5G MIMO antenna architecture is designed on 10-mil-thick Rogers 5870 substrate with dielectric constant (ε_r) of 2.33 ± 0.02 and loss tangent of 0.0012. Substrate of low relative permittivity was chosen in order to minimize surface wave modes. In addition, the low radiation efficiency is also the consequence of using the substrate of high dielectric constant and high loss tangent; 10-mil substrate is used as it is the optimal choice for achieving corner bending. Thin substrate was chosen in order to decrease cross polarization; 5-mil substrate is more flexible than the proposed substrate but requires additional scaffolding. Flexible substrates such as polyethylene terephthalate (PET) and polycarbonate have a high dielectric loss tangent, which will result in additional gain deterioration at mmWave frequencies.

7.5.1 4G LTE Antenna Design

Schematics of the proposed 4G LTE antenna integrated with mmWave 5G antenna module is depicted in Figures 7.17(a) and 7.17(b), with top and bottom planes, respectively. The proposed 4G LTE section of the antenna is electrically small, having the dimensions of 0.08 λ × 0.17 λ × 0.001 λ at 1.7 GHz. A rectangular slot of proper dimensions is etched on the ground plane, which is excited by a 50 Ω microstrip line on the other plane. The proposed slot antenna achieves resonance depending on the relative position of the

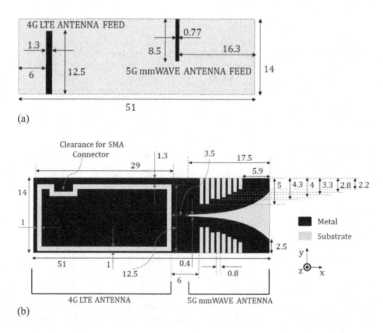

FIGURE 7.17
Schematics of the proposed integrated 4G LTE and mmWave 5G antenna (a) Top plane and (b) Bottom plane (all dimensions are in mm) [36].

microstrip feed, the width of the rectangular slot and the length of the feed line. The proposed antenna is optimized against these parameters, and the dimensions are given in Figure 7.17.

The proposed co-designed 4G LTE and mmWave 5G antenna is fabricated and the photograph is shown in Figures 7.18(a) and 7.18(b), with top and bottom planes, respectively. The simulated and measured input reflection coefficient of the proposed 4G LTE antenna is illustrated in Figure 7.19(a). Measured results were carried out using Agilent PNA E8364C. The measured impedance bandwidth of the proposed 4G LTE antenna is from 1.7 to 3 GHz. The proposed antenna is wideband, having a fractional bandwidth of 55%. The proposed prototype covers multiple LTE bands such as LTE1700 (1710–2170 MHz), LTE2300 (2300–2400 MHz) and LTE2500 (2500–2690 MHz), thus making carrier aggregation possible for achieving higher data rates. Discrepancies between simulated and measured data may be related to fabrication tolerances. Also, frequency shift may be caused by the inhomogeneity of the relative permittivity of the substrate.

The broadside gain and radiation efficiency plot of the proposed antenna are depicted in Figure 7.19(b). Gain of the proposed antenna at the operating LTE bands lies between 1.7 and 2.1 dBi, indicating high gain yield for the given electrical size. In addition to this, radiation efficiency at the LTE bands lies between 60% and 85%. The high efficiency of the proposed antenna is a result of the electrically thin substrate with a low dissipation factor. Simulated and measured radiation patterns of the proposed antenna in the YZ-plane (E-plane) are illustrated in Figures 7.20(a), 7.20(b) and 7.20(c). Measured results were carried out in an anechoic chamber. The proposed antenna achieves pattern integrity with dipole-like radiation patterns. Disparity between simulated and measured data results from poor absorptivity of the oblique incidence in the anechoic chamber.

7.5.2 mmWave 5G Antenna Design

A tapered slot antenna with end-fire gain is designed for mmWave 5G applications. Schematics of the proposed mmWave 5G antenna co-designed with 4G LTE antenna are shown in Figures 7.17(a) and 7.17(b), with top and ground planes, respectively. The proposed mmWave 5G section of the antenna has the dimensions of $1.75\ \lambda \times 1.4\ \lambda \times$

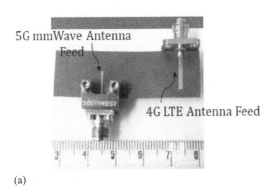
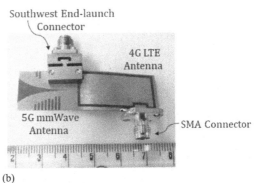

(a)   (b)

FIGURE 7.18
Photograph of the fabricated co-designed 4G LTE and mmWave 5G antenna (a) Top plane and (b) Ground plane [36].

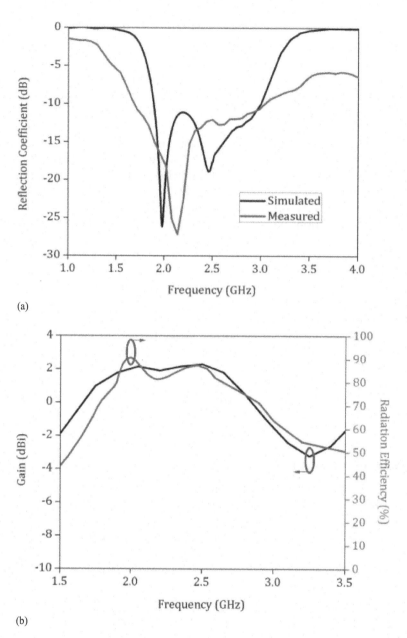

FIGURE 7.19
(a) Input reflection co-efficient and (b) Gain and radiation efficiency plots of the proposed 4G LTE antenna [36].

0.0254 λ at 28 GHz. The tapered slot antenna is a travelling wave antenna, a 50 Ω microstrip line is feeding the balun of a Vivaldi antenna, and microstrip to slot line transition is optimized for better impedance matching. The proposed antenna tapers exponentially in the ground plane. The antenna starts to radiate from the metallic taper until the electromagnetic wave has travelled around half of the wavelength. Corrugations are inserted in

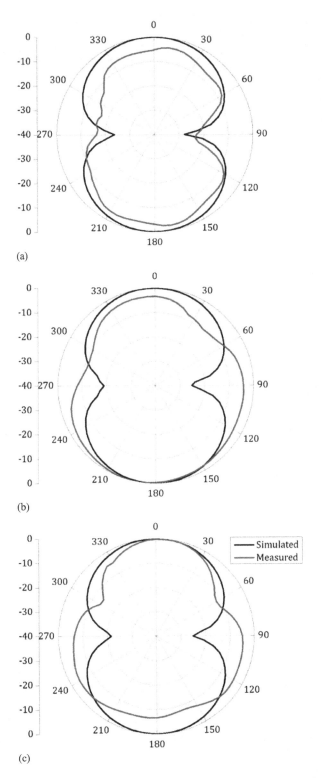

FIGURE 7.20
Simulated and measured radiation plots in the YZ-plane at (a) 1.8 GHz, (b) 2.3 GHz and (c) 2.6 GHz [36].

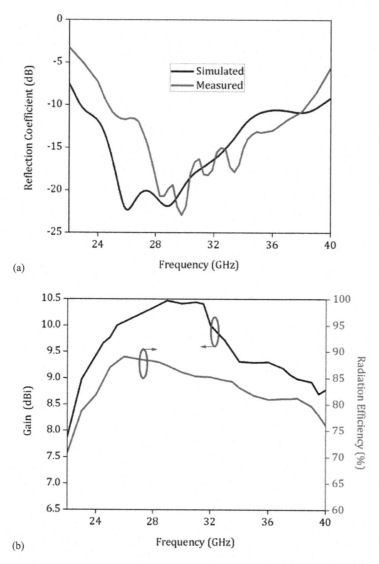

FIGURE 7.21
(a) Input reflection coefficient and (b) Gain and radiation efficiency plots of the proposed mmWave 5G antenna [36].

the proposed antenna to concentrate the E-field towards the main radiating aperture, thereby reducing back lobe. Thus cross polarization and side-lobe levels for the proposed antenna are reduced.

Photographs of the proposed mmWave 5G antenna are shown in Figures 7.18(a) and 7.18(b), with top and bottom planes, respectively. The simulated and measured input reflection coefficient of the proposed antenna is illustrated in Figure 7.21(a). The measured impedance bandwidth is high, ranging from 25 to 38 GHz with a fractional bandwidth of

41%. Discrepancies can be imputed to fabrication tolerances and deviation in port impedance offered by the end-launch connector. Gain of the proposed antenna lies between 9 and 10.5 dBi, indicating high gain for the available aperture, as depicted in Figure 7.21(b). Moreover, a 1-dB gain bandwidth of the proposed antenna is around 28%, with gain varying between 9.5 and 10.5 dBi. Standard gain transfer method is used for gain measurement using Keysight horn antennas. As path loss is high at mmWave frequencies, mobile phone antennas must possess high gain. Also, efficiency for the proposed antenna is high ranging between 80% and 88% as shown in Figure 7.21b.

The travelling wave phenomenon is realized as illustrated in Figure 7.22 with the travelling wave propagating at two different time instants. Simulated and measured radiation patterns of the proposed mmWave 5G antenna in the XY-plane at 28, 33 and 38 GHz are depicted in Figures 7.23(a), 7.23(b) and 7.23(c). The proposed antenna shows pattern integrity at all of the three given frequencies. The front-to-back ratio of the proposed tapered slot antenna is greater than 15 dB. Disparity between simulated and measured patterns is caused by alignment errors and the adapters used for pattern measurement.

7.5.3 Co-Designed Corner Bent 4G LTE and mmWave 5G MIMO Antennas

The proposed 4G LTE and mmWave 5G antennas are integrated with each other on the same 10-mil Rogers 5870 substrate and separated by a distance of 3.5 mm. Corner bending is carried out to decrease the effective radiating volume. It is also done to fit the antenna into the corner of a typical smartphone. The antenna's characteristics were observed and there was no significant variation in the input reflection coefficient or gain of either the 4G LTE or the mmWave 5G antenna.

For realizing a corner bent MIMO configuration, another co-designed 4G LTE and mmWave element is brought to the opposite corner of a smartphone, as illustrated in Figure 7.24. Moreover, isolation between the closely spaced integrated 4G LTE and mmWave 5G antennas is less than 30 dB, as shown in Figure 7.25(a). The two antenna elements are placed in such a way that orthogonal pattern diversity at mmWave frequencies is achieved, as depicted in Figure 7.25(b). Orthogonal pattern diversity results in the usage of a mobile phone in portrait as well as landscape mode. Table 7.3 illustrates the comparison between proposed 4G LTE and mmWave corner bent MIMO antenna design with other reported designs.

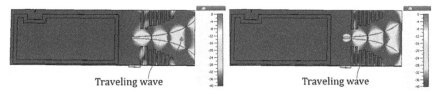

FIGURE 7.22
E-field plots for the proposed mmWave 5G antenna at (a) $t = 0$ and (b) $t = T/4$ [36].

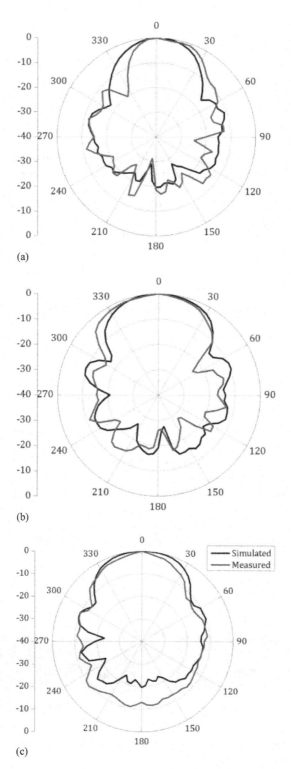

FIGURE 7.23
Simulated and measured radiation plots in XY-plane at (a) 28 GHz, (b) 33 GHz and (c) 38 GHz [36].

Co-Design of 4G LTE and mmWave 5G Antennas for Mobile Terminals   183

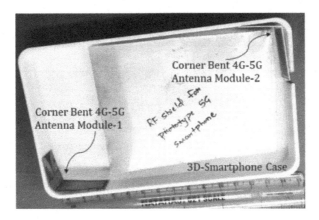

FIGURE 7.24
Integrated 4G LTE and mmWave 5G corner bent MIMO antenna design inside a typical mobile phone case [36].

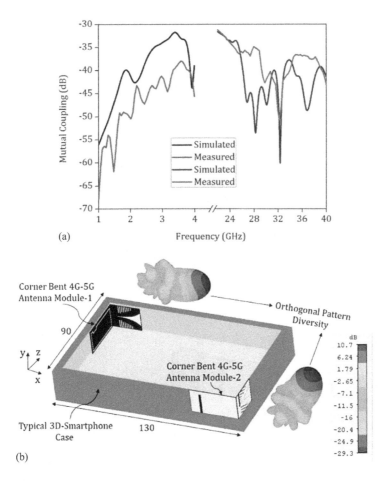

FIGURE 7.25
(a) Isolation plot of the co-designed 4G LTE and mmWave antenna element and (b) Orthogonal pattern diversity architecture with 3D patterns at 28 GHz [36].

TABLE 7.3

Comparison of the Proposed MIMO Antenna Design with other Reported Designs

Figures of Merit	Proposed Work	[2]	[4]
4G LTE Antenna			
Volume of Single Element	$30 \times 14 \times 0.254$ mm^3	$75 \times 8 \times 7$ mm^3	$9 \times 30 \times 0.965$ mm^3
Fractional Bandwidth (BW)	55% (−10 dB BW)	31%-Low Band, 44%-High Band (−6 dB BW)	30% (−6 dB BW)
Operating LTE Bands	LTE1900/2300/2500	LTE700/1900/2300/2500	LTE1900/2300/2500
Efficiency	60%–90%	50%–90%	50%–83%
MIMO	Yes	No	Yes
Corner Bent	Yes	No	No
mmWAVE 5G Antenna			
Volume of antenna	$17.5 \times 14 \times 0.254$ mm^3 (Single Element)	$23 \times 7 \times 4$ mm^3 (1 × 4 array)	$23.2 \times 8.3 \times 0.965$ mm^3 (2 × 4 array)
Impedance Bandwidth	25–38 GHz	25–30 GHz	26–28.4 GHz
Peak Realized Gain	10.5 dB	7 dB	8.2 dB
1-dB Gain Bandwidth	28%	Not Available	Not Available
Orthogonal Pattern Diversity	Yes	No	No
MIMO	Yes	No	No
Corner Bent	Yes	No	No

7.6 Case Study: Co-Design of 4G and 5G Antennas in a Smartphone

The proposed antennas reported in Sections 7.3 and 7.4 are integrated to investigate their characteristics. The schematics of the proposed integrated antenna system are shown in Figure 7.26.

The 4G LTE and mmWave 5G MIMO antennas are brought close to each other, separated by a distance of 0.2 mm. The antennas' characteristics were observed and there was

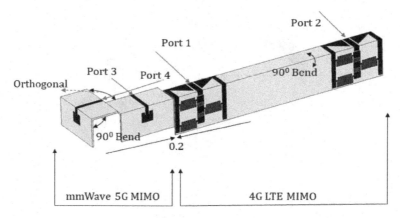

FIGURE 7.26
Schematics of integrated 4G LTE MIMO and mmWave 5G MIMO antenna [36].

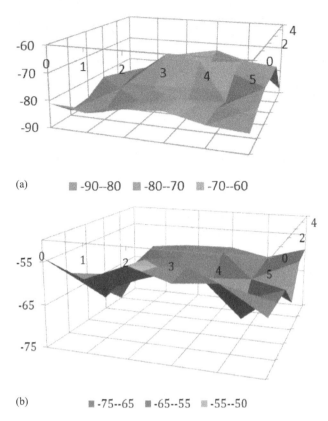

FIGURE 7.27
Received power profile for (a) 5G and (b) 4G [36].

no significant effect on the reflection coefficient or gain of either the 4G LTE or the mmWave 5G MIMO antenna. Further, isolation is less than 40 dB ($|S_{31}|$ or $|S_{13}|$) between two closely spaced 4G LTE/mmWave 5G antennas and less than 55 dB between the distant 4G/5G antennas. In order to illustrate the operation of the proposed co-design, a real-world deployment was performed, as can be seen from FigureS 7.27 (a) and 7.27(b).

7.7 Conclusion

In this chapter, a compact conformal integrated design for 4G LTE and mmWave 5G antennas has been proposed. The 4G LTE MIMO antenna system consists of two CRLH-based CPW-fed conformal antennas which relate to the class-7 4G LTE band. The proposed antenna is designed on a 10 mil thin substrate with electrically small dimensions of 20×20 mm². For the mmWave 5G MIMO antenna system, conformal orthogonally placed microstrip-fed antennas are proposed. The conformal mmWave 5G antenna system operates at 28 GHz with a forward gain of 9 dBi. 4G LTE and 5G mmWave MIMO antennas are integrated electrically close by 0.004λ to achieve minimal physical footprint. Real-world deployment was performed for both the 4G LTE and the mmWave 5G MIMO antenna systems. All the results validate that the proposed co-design is a good candidate for mobile terminals.

References

1. J. Lu and Y. Wang, "Planar small-size eight-band LTE/WWAN monopole antenna for tablet computers," *IEEE Transactions on Antennas and Propagation*, 62(8), 4372–4377, August 2014.
2. J. Kurvinen, H. Kähkönen, A. Lehtovuori, J. Ala-Laurinaho, and V. Viikari, "Co-designed mm-wave and LTE handset antennas," *IEEE Transactions on Antennas and Propagation*, 67(3), 1545–1553, March 2019.
3. M. S. Sharawi, M. Ikram, and A. Shamim, "A two concentric slot loop based connected array MIMO antenna system for 4G/5G terminals," *IEEE Transactions on Antennas and Propagation*, 65(12), 6679–6686, December 2017.
4. R. Hussain, A. T. Alreshaid, S. K. Podilchak, and M. S. Sharawi, "Compact 4G MIMO antenna integrated with a 5G array for current and future mobile handsets," *IET Microwave Antennas Propagation*, 11(2), 271–279, February 2017.
5. D. J. Chung, S. K. Bhattacharya, G. E. Ponchak, and J. Papapolymerou, "An 8×8 lightweight flexible multilayer antenna array," *2009 IEEE Antennas and Propagation Society International Symposium*, Charleston, SC, 1–4, 2009.
6. M.S. Sharawi, *Printed MIMO Antenna Engineering*, Norwood, MA, USA: Artech House, 2014.
7. S. F. Jilani and A. Alomainy, "Planar millimeter-wave antenna on low-cost flexible PET substrate for 5G applications," *2016 10th European Conference on Antennas and Propagation (EuCAP)*, Davos, 1–3, 2016.
8. K. Sarabandi, J. Oh, L. Pierce, K. Shivakumar, and S. Lingaiah, "Lightweight, conformal antennas for robotic flapping flyers," *IEEE Antennas and Propagation Magazine*, 56(6), 29–40, December 2014
9. H. Wong, K. K. So, K. B. Ng, K. M. Luk, C. H. Chan, and Q. Xue, "Virtually shorted patch antenna for circular polarization," *IEEE Antennas and Wireless Propagation Letters*, 9, 1213–1216, 2010.
10. D. Wang, H. Wong, and C. H. Chan, "Small patch antennas incorporated with a substrate integrated irregular ground," *IEEE Transactions on Antennas and Propagation*, 60(7), 3096–3103, July 2012
11. N. Amani and A. Jafargholi, "Zeroth-order and TM_{10} modes in one-unit cell CRLH mushroom resonator," *Antennas and Wireless Propagation Letters IEEE*, 14, 1396–1399, 2015
12. V. Rajasekhar Nuthakki and S. Dhamodharan, "Via-less CRLH-TL unit cells loaded compact and bandwidth-enhanced metamaterial based antennas," *AEU – International Journal of Electronics and Communications*, 80, 48–58, 2017.
13. C. Zhang, J. Gong, Y. Li, and Y. Wang, "Zeroth-order-mode circular microstrip antenna with patch-like radiation pattern," *Antennas and Wireless Propagation Letters IEEE*, 17(3), 446–449, 2018.
14. S. K. Sharma and R. K. Chaudhary, "A compact zeroth-order resonating wideband antenna with dual-band characteristics," *IEEE Antennas and Wireless Propagation Letters*, 14, 1670–1672, 2015
15. L. Si, W. Zhu, and H. Sun, "A compact, planar, and CPW-fed metamaterial-inspired dual-band antenna," *IEEE Antennas and Wireless Propagation Letters*, 12, 305–308, 2013
16. B. Bharathi and S. K. Koul, "Stripline-Like Transmission Lines for Microwave Integrated Circuits. New York: Wiley, 1989.
17. G. S. Karthikeya, M. P. Abegaonkar, and S. K. Koul, "CPW fed wideband corner bent antenna for 5G mobile terminals," *IEEE Access*, 7, 10967–10975, 2019
18. C. Lai, S. Chiu, H. Li, and S. Chen, "Zeroth-order resonator antennas using inductor-loaded and capacitor-loaded CPWs," *IEEE Transactions on Antennas and Propagation*, 59(9), 3448–3453, September 2011
19. N. Amani, M. Kamyab, A. Jafargholi, A. Hosseinbeig, and J. S. Meiguni, "Compact tri-band metamaterial-inspired antenna based on CRLH resonant structures," *Electronics Letters*, 50(12), 847–848, 2014

20. S. C. Chiu, C. P. Lai, and S. Y. Chen, "Compact CRLH CPW antennas using novel termination circuits for dual-band operation at zeroth-order series and shunt resonances," *Antennas and Propagation IEEE Transactions on*, 61(3), 1071–1080, 2013.
21. A. Lai, K. M. K. H. Leong, and T. Itoh, "Infinite wavelength resonant antennas with monopolar radiation pattern based on periodic structures," *IEEE Transactions on Antennas and Propagation*, 55(3), 868–876, March 2007.
22. J. Zhu and G. V. Eleftheriades, "A compact transmission-line metamaterial antenna with extended bandwidth," *IEEE Antennas and Wireless Propagation Letters*, 8, 295–298, 2009.
23. M. A. Abdalla, "A high selective filtering small size/dual band antenna using hybrid terminated modified CRLH cell," *Microwave and Optical Technology Letters*, 59(7), 1680–1686, 2017.
24. A. Mehdipour, T. A. Denidni, and A. R. Sebak, "Multi-band miniaturized antenna loaded by ZOR and CSRR metamaterial structures with monopolar radiation pattern," *IEEE Transactions on Antennas and Propagation*, 62(2), 555–562, 2014.
25. B. Yang, Z. Yu, Y. Dong, J. Zhou, and W. Hong, "Compact tapered slot antenna array for 5G millimeter-wave massive MIMO systems," *IEEE Transactions on Antennas and Propagation*, 65(12), 6721–6727, December 2017
26. S. X. Ta, H. Choo, and I. Park, "Broadband printed-dipole antenna and its arrays for 5G applications," *IEEE Antennas and Wireless Propagation Letters*, 16, 2183–2186, 2017.
27. G. S. Reddy, A. Kamma, S. Kharche, J. Mukherjee, and S. K. Mishra, "Cross-configured directional UWB antennas for multidirectional pattern diversity characteristics," *IEEE Transactions on Antennas and Propagation*, 63(2), 853–858, February 2015
28. B. Zhou, H. Li, X. Zou, and T.-J. Cui, "Broadband and high-gain planar vivaldi antennas based on inhomogeneous anisotropic zero-index metamaterials," *Progress in Electromagnetics Research*, 120, 235–247, 2011.
29. F. Liu, J. Guo, L. Zhao, X. Shen, and Y. Yin, "A meta-surface decoupling method for two linear polarized antenna array in Sub-6 GHz base station applications," *IEEE Access*, 7, 2759–2768, 2019
30. Y. Sharma, D. Sarkar, K. Saurav, and K. V. Srivastava, "Three-element MIMO antenna system with pattern and polarization diversity for WLAN applications," *IEEE Antennas and Wireless Propagation Letters*, 16, 1163–1166, 2017
31. A. Dadgarpour, B. Zarghooni, B. S. Virdee, and T. A. Denidni, One- and two-dimensional beam-switching antenna for millimeter-wave MIMO applications," *IEEE Transactions on Antennas and Propagation*, 64(2), 564–573, February 2016
32. Z. Briqech, A. Sebak, and T. A. Denidni, "Wide-scan MSC-AFTSA array-fed grooved spherical lens antenna for millimeter-wave MIMO applications," *IEEE Transactions on Antennas and Propagation*, 64(7), 2971–2980, July 2016
33. M. Sun, Z. N. Chen, and X. Qing, "Gain enhancement of 60-GHz antipodal tapered slot antenna using zero-index metamaterial," *IEEE Transactions on Antennas and Propagation*, 61(4), 1741–1746, April 2013
34. Z. Wani, M. P. Abegaonkar, and S. K. Koul, "Millimeter-wave antenna with wide-scan angle radiation characteristics for MIMO applications," *International Journal of RF and Microwave Computer Aided Engineering*, 29, e21564, 2019.
35. M. Idrees Magray, G. S. Karthikeya, K. Muzaffar, and S. K. Koul, "Compact co-design of conformal 4G LTE and mmWave 5G antennas for mobile terminals," *IETE Journal of Research*, 2019. doi:10.1080/03772063.2019.1690593.
36. M. Idrees Magray, G. S. Karthikeya, K. Muzaffar, and S. K. Koul, "Corner bent integrated design of 4G LTE and mmWave 5G antennas for mobile terminals," *Progress in Electromagnetics Research M*, 84, 167–175, 2019

8

Corner Bent Phased Array for 5G Mobile Terminals

8.1 Introduction

Phased arrays are one of the most advertised antenna designs for millimeter wave 5G cellular devices, for numerous reasons. Since the higher path loss must be mitigated to achieve a reasonable link budget, it is important to design high gain antennas on the transceivers integrated with the mobile terminal. The problem is unique in the context of 5G mobile terminals, as previous commercial systems used electrically small antennas with poor gain. The simplest solution to increase gain of the antenna under consideration is to increase its effective radiating aperture [1,2]. One of the strategies to increase the radiating aperture would be to design a reflector-based antenna that can easily radiate a narrow beam with high gain, but the design would be bulky and inconvenient to integrate with the mobile terminal. It must also be noted that incorporating high gain reflector-based antennas on the base station would lead to poor angular coverage in spite of the reasonable link budget achieved. Several topologies of phased arrays have been proposed in recent years [3–10].

The other alternative is to design pattern diversity or beam switching antenna module, as demonstrated in the earlier chapters. Even though a pattern diversity module could be designed to serve for both landscape and portrait modes in a smartphone, the gain achievable from these designs is limited by the available physical footprint. Hence a phased array design might serve as an alternative option to design high gain antennas for mobile terminals with specified beam scanning and gain specifications. The width of the smartphone could be used to increase the number of elements of the radiators, consequently leading to gain enhancement. If phase shifters and their corresponding controller module could be developed as a CMOS chip then the transceiver with the phased array could be fit inside a mobile terminal.

The beam switching module for the base station presented in Chapter 6 offers a decent gain and mutual coupling, but the gain is in the range of 8–9 dBi across the band. The same radiating aperture phased array also could be designed with a higher degree of freedom in terms of beam scanning. Hence phased array could be an alternative for base station antenna designs. It must be noted that the phase-shifter back-end electronics complexity would be increased in the phased array design. The amount of effective power radiated could also be dynamically controlled in the case of a phased array design.

8.2 Phased Array Designs for mmWave Frequencies

A typical phased array system consists of identical antenna elements separated by a predetermined distance and fed by individual ports, with the individual ports' amplitude and offset phase shift deciding the location of the beam. By altering the relative phase difference between the ports, beam scanning could be obtained. The beam could be scanned within a specific angle because of the scanning loss phenomenon. In order to understand the nuances of phased array designs at 28 GHz, a generic 1 × 4 phased array is demonstrated in this section.

A standard inset-fed patch antenna is designed to operate at 28 GHz with 10% impedance bandwidth on 20 mil Nelco NY220 substrate, as illustrated in Figure 8.1. The radiator is placed 10 mm away from the feed plane to minimize the effects of the electrically large end-launch connector. But this design constraint increases the insertion loss caused by the feeding structure. The insertion loss because of the transmission line feeding the radiator is a crucial parameter to be considered when designing phased arrays with more than eight elements. As the number of elements increase, the transmission loss caused by the feeding structure would lead to a significant deterioration in the achievable gain. Hence, electrically thin substrates with very low dielectric loss must be the choice of substrate when designing high gain multi-port phased arrays on PCB.

The other alternative would be to use a waveguide-based feeding structure such as a substrate integrated waveguide (SIW) or a ridge-gap waveguide. However, the waveguide-based antennas need an additional transition to be incorporated into the feed-line to act as an impedance transformer from the waveguide TEM mode to the antenna. In this example, since the design is PCB based, no additional transition is required.

Figure 8.1 also illustrates the 1 × 4 phased array of patch antennas. The number of elements would primarily decide the gain and beamwidth in the H-plane. As the number of elements increase, so also does the gain. For example, the gain of 4 elements would be 3 dB higher than the gain of 2 elements, which means that as the number of elements is doubled, the gain is also expected to double in a linear scale, translating to 3 dB in a logarithmic scale. As a general rule of thumb, if X dBi is the element gain, then a phased array of 2^N elements is expected to have a gain of (X + 3N) dBi in the boresight, which serves as a

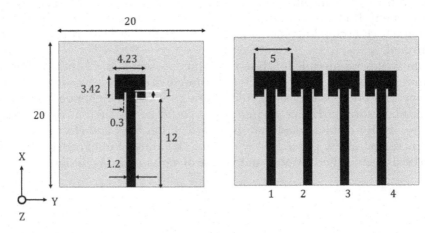

FIGURE 8.1
Schematic of the inset fed patch antenna and its 1 × 4 array.

rough estimate to gauge the number of elements for the desired gain. The estimated gain of (X + 3N) dBi would be feasible only if the elements are impedance matched post-integration, and mutual coupling is minimal, which is never the case in practical designs.

In this example, the antenna elements are separated by 5 mm, which is close to half a wavelength at 28 GHz. The separation is also an important parameter to be considered when designing phased arrays; if the separation is closer than half a wavelength then the beamforming would be inadequate because of the higher mutual coupling. If the elements are placed electrically further away from each other, grating lobes would appear when the ports are excited. Therefore the designer must choose the amount of separation to satisfy both the gain and mutual coupling requirements. The separation is usually fixed at a half wavelength at the centre frequency of operation. The amplitude and phase inputs to the individual ports decide the beamforming.

The broadside beam scanning of the 1 × 4 phased array is shown in Figure 8.2. The individual ports are pumped with equal power without any amplitude distribution. The ports are also fed with a progressive phase shift according to scan angle Equations (8.1) and (8.2), where 'n' is the nth port of the phased array, 'd' is the separation between the elements, and θ_{sa} is the angle at which the resultant beam would be oriented; λ is the free space wavelength computed at the centre frequency of operation. For example, if the beam is scanned at 30°, then $\theta_{sa} = 30°$, $d = \lambda/2$ hence, $\varphi_n = n\pi \sin(30°) = (n\pi)/2$

$$\varphi_n = nkd\sin\theta_{sa} \tag{8.1}$$

$$k = (2\pi)/(\lambda) \tag{8.2}$$

Table 8.1 displays the individual amplitude and phase of the corresponding ports to illustrate the beam scanning principle. The beam scanning for various beam scan angles is illustrated in Figure 8.2; it is apparent that the beamwidth increases as the beam is scanned

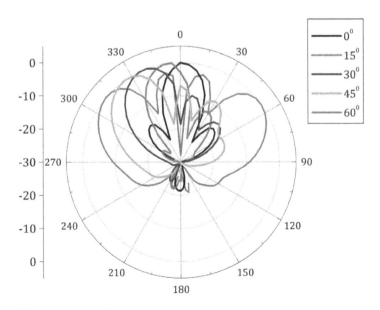

FIGURE 8.2
Broadside beam scanning of the 1 × 4 array.

TABLE 8.1
Amplitude and Phase Feeding for 30° Beam Scan

Port No.	Amplitude	Phase
1	1	0°
2	1	90°
3	1	180°
4	1	270°

away from the boresight leading to deterioration in gain, termed as scanning loss. It is also clear that the beam is almost unusable when it is scanned at 60° or higher because of poor beamforming.

8.3 Need for Corner Bent Phased Array

Most of the reported designs based on millimeter wave phased array are planar in nature, which would occupy a large physical footprint in addition to the wastage of radiated power towards the user. Statistically speaking, users would hold the smartphone at an angle of 30° with respect to the horizontal axis. In this situation, if the broadside radiating array is integrated with the panel, the radiation would be targeted towards the user and this would naturally cause a signal blockage as the human body offers a heavy attenuation (30–40 dB) to the millimeter wave signal. The situation is illustrated clearly in Figure 8.3.

To resolve this issue, an end-fire radiating array could be installed in the panel, such as a printed dipole array, but the physical footprint of the antenna would still be high, reducing the real-estate on the phone for the motherboard and other critical components. Hence the solution to designing a high gain antenna module with beam scanning and minimal footprint is to develop a corner bent broadside array that would radiate away from the user without causing any signal outage. Figure 8.4 gives a clear insight into this design criterion.

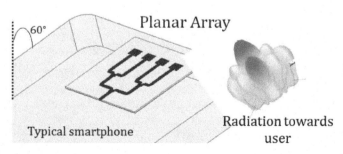

FIGURE 8.3
Radiation in a real-time scenario for a planar array.

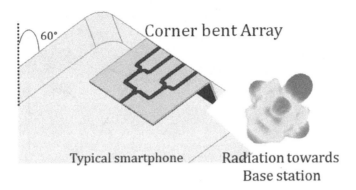

FIGURE 8.4
Radiation caused by corner bent array on a smartphone.

If the front-to-back ratio of the corner bent array is more than, say, 20 dB, it would mean that significant power is directed towards the base station as against the person using the smartphone. The phase shifters and controllers are not visualized in the design example.

To design the corner bent phased array it is essential to design the antenna on a flexible substrate similar to the antennas designed in section 3.3. The insertion loss is not a critical parameter for stand-alone antennas, or even a diversity module with two or three ports, but it is critical when designing a four-element array. As the number of radiating elements increase, the number of feed-lines and the power dividers would also increase, leading to additional insertion loss and consequently leading to degradation in the boresight gain.

To characterize the relatively low loss polycarbonate substrate, a 50 Ω transmission line is fabricated on a 500 μm substrate, as depicted in the photograph in Figure 8.5, the fabrication of which is identical to the process explained in section 3.3. The length of the transmission line is 50 mm, which translates to 5λ. The insertion loss of the planar transmission line

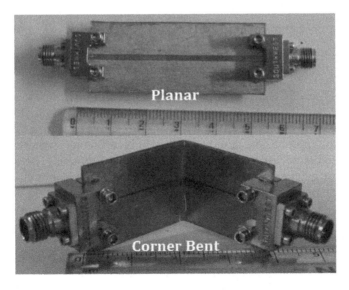

FIGURE 8.5
Photographs of the transmission lines on polycarbonate substrate.

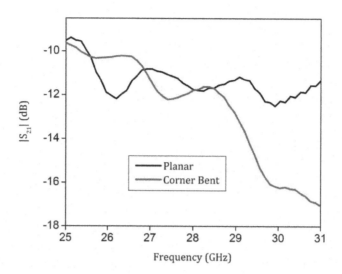

FIGURE 8.6
Measured insertion loss of planar versus corner bent transmission lines.

is illustrated in Figure 8.6, indicating a loss of close to 11 dB at 28 GHz. Since the antenna would be designed for corner bending, the transmission line would also be corner bent in the centre, as can be seen from Figure 8.5, and its corresponding insertion loss is also shown in Figure 8.6, indicating a quite close match to its planar counterpart. The loss increases at higher frequencies may be caused by additional radiation from the 90° bending. It is inferred that the bending loss is negligible for the substrate at these frequencies, hence a corner bent array could be designed with this substrate provided the bandwidth is limited to 27–29 GHz for minimal insertion loss resulting from feeding.

8.4 Corner Bent Phased Array on Polycarbonate

To design a phased array it is important to first characterize the elements' characteristics. The antenna element must have a minimal footprint with maximum forward gain achievable. The beam of the single element should be hemispherical, so that when used in the context of a phased array, a low side-lobe main beam could be formed. In this design, the corner bent patch antenna designed on the polycarbonate substrate introduced in section 3.3 is investigated for its suitability as an array element. The schematic and the radiation patterns are depicted in Figure 8.7 and Figure 8.8, respectively.

It is clear that the forward patterns of the antenna are almost hemispherical, with a high front-to-back ratio of close to 20 dB across the band. It must also be noted that the pattern integrity for the 10% impedance bandwidth is also high, which would translate to a decent beamforming when integrated in an array.

The forward gain is 7 dBi, as illustrated in Figure 8.9, for reasons explained in section 3.3. The gain could have been increased with a lower loss substrate, but the insertion loss post bending would in that case be higher. An electrically thinner substrate would also have given higher gain, but bending would create mechanical stress to the element. Also,

Corner Bent Phased Array for 5G Mobile Terminals

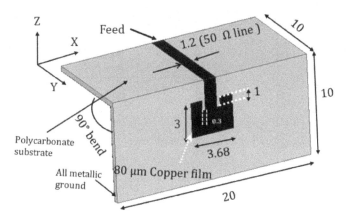

FIGURE 8.7
Schematic for corner bent inset fed patch antenna.

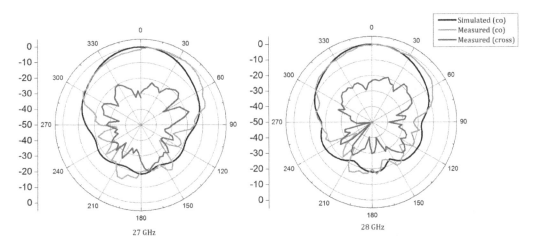

FIGURE 8.8
Forward H-plane patterns of the corner bent antenna.

mechanical sturdiness is an important design constraint when designing arrays, as phased arrays tend to occupy a larger footprint, and hence require greater mechanical strength.

The 1 × 4 antenna array is illustrated in Figure 8.10, where the corner bent schematic is also depicted. The fabricated prototype of the flexible array is shown with the end-launch connector in Figure 8.11. Ideally, four different ports have to be designed with individual feeding. This strategy would require four electrically large end-launch connectors along with four expensive transceiver modules.

It is a common practice in the antenna community to demonstrate the characteristics of the antenna array with the power divider as seen in the schematics. The feed of the individual antennas are 50 Ω, but the incoming feed line also is designed as a 50 Ω line, thus a two-stage power divider is designed for convenience.

Consider the impedances of antennas 1 and 2, which offer 50 Ω each, and since both of these are in shunt as seen from the feed line, the effective net impedance offered by the combination is 25 Ω, which has to be matched to the 50 Ω line of the left wing of the first

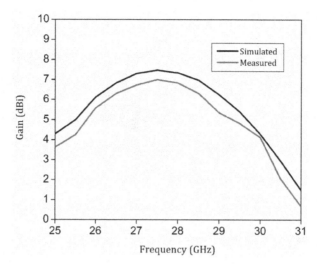

FIGURE 8.9
Forward gain of the stand-alone antenna element.

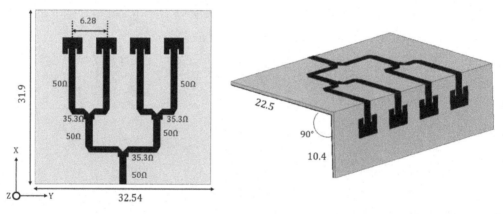

FIGURE 8.10
Schematic of 1 × 4 antenna array with 0° beam scanning (planar and corner bent).

power divider. A quarter-wave transformer of impedance 35.35 Ω and 90° length is inserted between the first port of the first power divider and the feed lines of the antenna elements. The same strategy is followed for antenna elements 3 and 4, to achieve impedance matching. The separation between the elements is optimized for 6.28 mm for the reasons mentioned in section 8.2.

The mutual coupling in the power-divider-based array would naturally be higher than the individual ports' feeding network. It must be noted that the isolation between the fingers of the power divider is not remarkable, as no Wilkinson network is used. If a 100 Ω resistor was used between the antenna elements, isolation between the elements would improve, leading to higher gain in boresight. But resistors would be adding a series inductance post soldering, leading to detuning of the antenna post integration, especially at frequencies beyond 20 GHz, hence resistor design was avoided.

Corner Bent Phased Array for 5G Mobile Terminals 197

FIGURE 8.11
Photograph of the corner bent array.

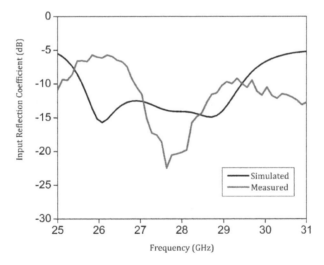

FIGURE 8.12
Input $|S_{11}|$ of the corner bent array.

The simulated and measured input reflection coefficients are illustrated in Figure 8.12. The impedance bandwidth is decided primarily by the feed network feeding the elements and the impedance bandwidth of the individual element itself. It is challenging to design a wider bandwidth in the broadside orientation compared to an end-fire antenna.

The radiation patterns of the boresight scanned phased array at 28 and 29 GHz are shown in Figure 8.13(a) and Figure 8.13(b). The front-to-back ratio is also close to 20 dB as a result of the electrically large ground plane. The side-lobes are caused by scattering of the feed network and the corner bending. It must also be noted that the array was fed with

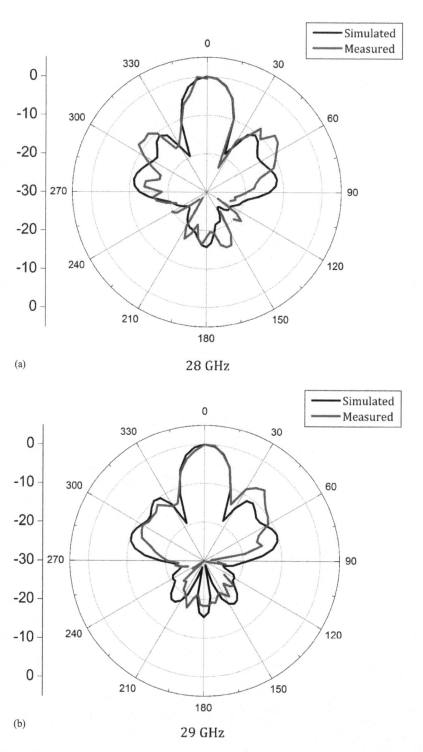

FIGURE 8.13
(a) H-plane pattern at 28 GHz and (b) H-plane pattern at 29 GHz.

equal amplitude and phase, hence leading to relatively higher side-lobe level. The forward gain of the array is illustrated in Figure 8.14; the boresight gain is 10 dBi at 28 GHz as against 7 dBi of the single element. According to the gain estimation formula mentioned in section 8.2, the gain of the array with power divider should have been 13 dBi, the discrepancy between the estimate and actual design being caused by low isolation between the elements and the high insertion loss of the feeding network.

To demonstrate the beam scanning principle, a 30° beam scanned array is designed as shown in Figure 8.15, along with its corresponding pattern in Figure 8.16. The progressive phase shift indicated in Table 8.1 is implemented as delay lines in the feeding lines of the respective antennas.

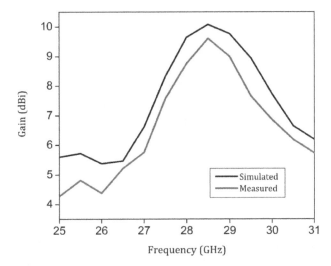

FIGURE 8.14
Forward gain of the corner bent array.

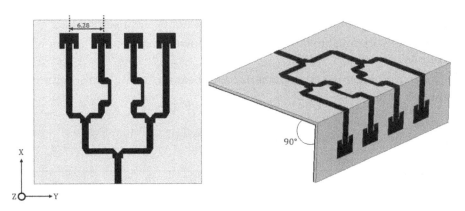

FIGURE 8.15
Schematic of 1 × 4 antenna array with 30° beam scanning (planar and corner bent).

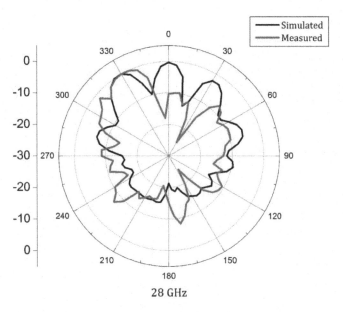

FIGURE 8.16
Forward pattern for the 30° beam scanned corner bent array.

8.5 Design Guidelines for a Phased Array at Ka Band

1. Antenna element design must operate in the specified band with a hemispherical beam and maximum gain for the available aperture. The antenna must be matched to the standard 50 Ω as this would be easier for scaling up the design in an actual phased array.
2. The specified gain decides the number of elements required to achieve that gain. Simple calculations must give an estimate regarding the number of elements; additional elements might be required to compensate for the insertion loss.
3. The feed network must to be designed with power dividers and additional matching circuits as necessary.
4. The beam forming must be characterized for beam scanning and scanning loss across frequency of interest.

8.6 Conclusion

In this chapter, the principle of phased arrays was introduced, followed by design criteria for phased array design specifically at mmWave 5G frequencies. The principle for beam scanning with respect to a generic 1 × 4 inset fed patch antenna array was also discussed.

The need for a corner bent phased array topology was also discussed with justification. A corner bent phased array on a flexible substrate was demonstrated with a power divider network. The guidelines for designing a phased array for Ka band was also discussed.

References

1. R. J. Mailloux, *Phased Array Antenna Handbook*, Norwood, MA, USA, Artech House, 1994.
2. R. C. Hansen, *Phased Array Antennas*, New York, Wiley, 1998.
3. W. Hong, K. Baek, Y. Lee, Y. Kim, and S. Ko, "Study and prototyping of practically large-scale mmWave antenna systems for 5G cellular devices," *IEEE Communications Magazine*, 52(9), 63–69, September 2014.
4. S. F. Jilani and A. Alomainy, "Planar millimeter-wave antenna on low-cost flexible PET substrate for 5G applications," *2016 10th European Conference on Antennas and Propagation (EuCAP)*, Davos, 1–3, 2016.
5. S. F. Jilani, M. O. Munoz, Q. H. Abbasi and A. Alomainy, "Millimeter-wave liquid crystal polymer based conformal antenna array for 5G applications," *IEEE Antennas and Wireless Propagation Letters*, 18(1), 84–88, January 2019.
6. S. X. Ta, H. Choo, and I. Park, "Broadband printed-dipole antenna and its arrays for 5G applications," *IEEE Antennas and Wireless Propagation Letters*, 16, 2183–2186, 2017.
7. S. Zhu, H. Liu, Z. Chen, and P. Wen, "A compact gain-enhanced vivaldi antenna array with suppressed mutual coupling for 5G mmWave application," *IEEE Antennas and Wireless Propagation Letters*, 17(5), 776–779, May 2018.
8. B. Yang, Z. Yu, Y. Dong, J. Zhou, and W. Hong, "Compact tapered slot antenna array for 5G millimeter-wave massive MIMO systems," *IEEE Transactions on Antennas and Propagation*, 65(12), 6721–6727, December 2017.
9. M. Sun, Z. N. Chen, and X. Qing, "Gain enhancement of 60-GHz antipodal tapered slot antenna using zero-index metamaterial," *IEEE Transactions on Antennas and Propagation*, 61(4), 1741–1746, April 2013.
10. R. A. Alhalabi and G. M. Rebeiz, "High-efficiency angled-dipole antennas for millimeter-wave phased array applications," *IEEE Transactions on Antennas and Propagation*, 56(10), 3136–3142, October 2008.

9
Fabrication and Measurement Challenges at mmWaves

9.1 Introduction

Fabrication of the antennas is one of the most crucial part of the entire design process. The challenges offered for characterizing antennas in the Ka band is unique because of the physical size of the antenna and the various aspects involved in pattern measurements. It must be noted that antenna measurement in the lower frequencies (1–10 GHz) is straightforward as the sources of error are minimal and well known [1,2]. Insights on issues with millimeter wave antenna measurements are discussed in [3–6]. In this chapter, issues regarding fabrication and measurement is discussed in detail.

9.2 Fabrication Process and Associated Tolerances

PCB-based antennas are fabricated using industry standard procedures. The fabrication process is discussed below, along with sources of error.

1. First, after choosing the substrate it is highly recommended to measure the dielectric constant and loss tangent at the frequency of operation, since manufacturers typically specify parameters of the substrate at 10 GHz so these parameters might not be valid at 28 GHz [7–12]. The thickness of the substrate must also be measured. The double-sided copper-clad substrate of sufficient size is cleaned using a powerful detergent, followed by rinsing with acetone. Typically, these substrates are shipped from companies abroad and over the time spent in transit accumulate dust and detritus. Copper cladding on the substrate must be as clean as possible for the copper to be able to react adequately to the fabrication process. The substrate must be cleaned on both the sides, since typically antenna designs will be double-sided. The substrate is then blow-dried to remove the remaining water droplets.

2. The substrate is taken into a developing room lit with yellow light, where a thin layer of photoresist is applied to the substrate. The photoresist-coated substrate is then pasted on to a spinner and revolved for about 1–2 min at a uniform speed of 500 rpm. Spinning ensures a uniform coating of photoresist on the substrate. Care must be taken to remove any residual adhesive from the substrate. After spinning, the

substrate is blow-dried to harden the photoresist layer on both the sides of the substrate. The photoresist application is the most important step in this process, since thick layers would lead to difficulties in dissolving the photoresist during the development stage, and layers that are too thin would lead to inadequate pattern formation during the development stage. After a few trials any researcher should be able to pick up this skill.

3. The mask of the pattern must be printed beforehand. A simulation software package such as Ansys HFSS would generally be able to export the design file in .dxf format, which could be sent to print the mask. Care must be taken to add additional space around the antenna for the operator to handle the substrate. Also, the resolution of the mask must be known a priori, which means that the resolution of the printer must be carefully noted before the design is sent. For example, the masks used in the antennas presented in this book had a resolution of 100 µm, which means that the smallest dimension of the structure in the geometry of the antenna cannot be lower than this dimension.

4. The mask is cut and pasted appropriately to the photoresist-coated substrate. The alignment of the mask and the substrate is as important as the coating itself. The alignment might be a challenging task in double-sided designs. It would be a clever move to include intentional alignment markers in the mask which could be cut off post-etching. The alignment also decides the accuracy of the input impedance.

5. The photoresist-coated substrate aligned with the mask is placed inside an ultraviolet (UV) exposure unit for 1–2 min, the duration of the exposure time being decided by the complexity of the geometry. For example, if the design is composed of physically thin lines of sub-wavelength structures, then 1 min would be sufficient. On the other hand, if the design is 10 Ω transmission line on a 20 mil substrate, then UV exposure for 2 min would be necessary.

6. After the UV exposure, the PR hardened substrate must be baked in an electric oven for 30–60 s. If an oven is unavailable, hot air drying could be done instead. Exposure to hot air for too long might cause additional hardening of the PR, leading to poorly developed patterns.

7. The mask must be removed and the PR-coated substrate must be developed in a solution of trichloroethylene followed by rinsing in acetone. This is the second most important step of the fabrication process. The developing must be performed carefully for 30–45 s. The optimal developing technique depends on the skill and experience of the researcher.

8. The developed substrate is hot-air dried and taken out of the developing room. Light-sensitive ink is then poured on to the substrate to identify the pattern. The pattern must resemble the geometry as seen in the mask. If this is not the case, the poorly developed substrate must be cleansed with acetone and steps 2–8 repeated.

9. The dried and inked substrate is now ready for chemical etching. A concentrated solution of ferric chloride is poured repeatedly on to the substrate to achieve the final prototype. If the geometry is composed of physically thin lines, then diluted etching solution may be used. Once the design is ready it must be blow-dried, followed by inspection under a microscope to search for any defects. If the defects are minor, they may be ignored, but if larger, the design must be re-fabricated. If the feed lines have a discontinuity, impedance matching would be poor. If there are errors in the aperture fabrication, the measured gain and patterns would be hampered.

10. Once the design is ready, it is important to smooth the portion of the transmission line with sandpaper to ensure a better connection with the connector. A multimeter might be used to check the connectivity of the connector mounting.

To fully appreciate the consequences of errors in fabrication, a case study is presented here. Consider the tapered slot antenna presented in section 5.2.1, as shown in Figure 9.1. The design is realized on Nelco NY9220 substrate with 20 mil thickness. The manufacturer states that the dielectric constant is 2.2 ± 0.02 at 10 GHz. The antenna resonates at 28 GHz if the dielectric constant is assumed to be 2.2, but the antenna detunes to 29 GHz when ε_r is 2.02. The antenna's resonance shifts to 27 GHz when ε_r is 2.31, as observed in Figure 9.2.

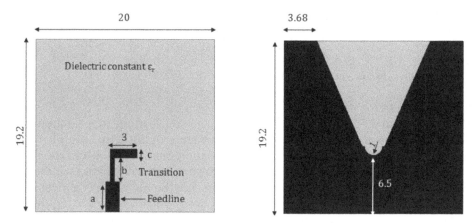

FIGURE 9.1
Schematic of the tapered slot antenna for tolerance analysis.

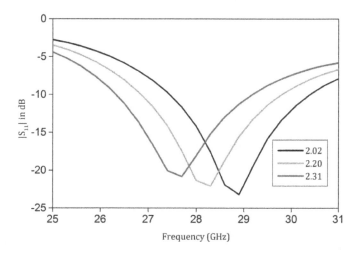

FIGURE 9.2
$|S_{11}|$ variation with variation in the dielectric constant of the substrate.

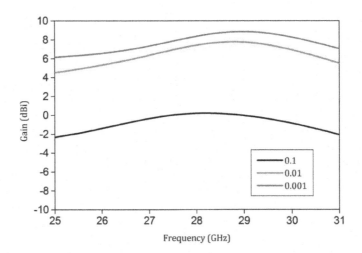

FIGURE 9.3
End-fire forward gain variation with dielectric loss tangent.

This translates to an 8.5% shift in the dielectric constant leading to a frequency shift of 3.5%. Researchers are advised that the dielectric constant of the actual sample might vary by a maximum of 5%. A tolerance analysis in simulation might be used as a rough estimate for the measured results. The forward end-fire gain variation with changes in the dielectric loss tangent of the substrate is illustrated in Figure 9.3, the radiation is almost hampered when the loss tangent is 0.1, indicating to heavily lossy substrate indicating that most of the energy is converted to surface waves. Gains for loss tangents 0.01 and 0.001 are also depicted in the graph of Figure 9.3.

To further understand the influence of manufacturing defects on the impedance characteristics of the antenna, the feedline length ('a' in Figure 9.1) and the transformer length ('b' in Figure 9.1) are assumed to be variable to see its influence on $|S_{11}|$, as demonstrated in Figures 9.4(a) and 9.4(b). It is clear that the errors in fabrication leading to variation in the actual dimensions of the fabricated prototype have minimal influence on the input impedance of the tapered slot antenna. The tolerance analysis presented here could be used to analyse the effects of faulty fabrication on the characteristics of the antenna.

9.3 S-parameter Measurements

Typically, end-launch connectors are used for scattering parameter measurements up to 40 GHz due to their low insertion loss and low radiation from the connector ensemble. Even though SMA connectors are characterized up to 26.5 GHz, beyond this frequency the connector would add additional insertion loss. But the main problem with conventional SMA connectors is that the width of the 50 Ω transmission line on an electrically thin substrate would be in the range of 0.5–1.2 mm, but the width of the trace of the SMA connector is almost 1.5 mm, hence creating an additional impedance mismatch. It must also be noted that the end-launch connectors might give good S-parameter measurements but are

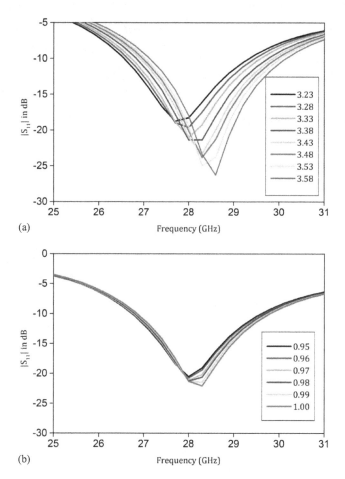

FIGURE 9.4
(a) $|S_{11}|$ variation with the length of the feed line and (b) $|S_{11}|$ variation with the length of the transition.

electrically large, leading to poor radiation pattern measurements. Hence it is recommended that the primary radiator of the antenna must be 10–15 mm away from the feeding point.

To measure mutual coupling between the antennas in a multi-port system, the spacing between the feed lines must be designed to accommodate the end-launch connectors.

9.4 Pattern Measurements and Sources of Error

Radiation pattern measurements are challenging to perform with mmWaves because of the relatively small physical footprint of the antenna with respect to the setup. The typical setup for pattern measurement is depicted in Figure 9.5. The Ka band horn operates in the 26.5–40 GHz range with a reasonably stable gain across the band. To ensure the operation of the standard horn, it is important also to benchmark the reference antenna by measuring the radiation pattern of the standard horn using another identical horn as the receiver. The mounting

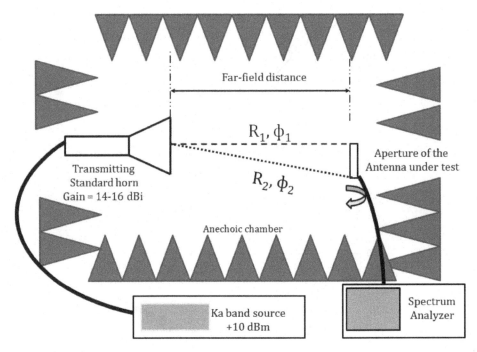

FIGURE 9.5
Setup for radiation pattern measurement.

for the physically small horn antenna must be designed with care, since the adjacent cables and adaptors must not create a high scattering environment for the measurement.

The horn is connected to the standard Ka band signal source which transmits pure sinusoid of designated power and frequency. Typically, the power transmitted would be set to the highest possible value of the source as the path loss is relatively higher. The power setting must also make sure that the horn is driven within its power ratings.

The antenna under test is aligned with respect to the primary beam of the horn antenna at far-field distance. A laser pointer or alignment device could be used to ensure the maximum power reception. It is highly recommended that the researcher always takes multiple measurements to optimize the alignment. The maximum power reception could be used for the rest of the measurements in other orthogonal planes or frequencies. The antenna under test is connected to the spectrum analyser placed outside the anechoic chamber.

The received power must be well above the noise floor of the spectrum analyser. For example, imagine the horn gain is 14 dBi at 28 GHz and the antenna under test has an estimated gain of 8 dBi and placed at 1m from the transmitting horn antenna. For this scenario, the free space path loss would be 40 dB and the cable loss approximately 30 dB at 28 GHz. This means that the effective power transmitted to the horn terminals is 10 dBm–30 dB = −20 dBm.

This power is increased by 14 dBi by the horn, leading to −20 dBm + 14 dBi = −6 dBm. This is the power at the aperture of the horn. The incoming power before the aperture of the antenna under test is −6dBm −40 dB because of the free space path loss leading to −46 dBm, but this power would be raised by 8 dBi, leading to −38 dBm, which would be the power read by the spectrum analyser.

This value is well above the noise floor of −110 dBm. It is important to note that this value of power is valid when the antenna is being measured in the co-polarized mode, that too is in the main lobe of radiation. If the antenna is being measured in the back lobe or in the cross-polarized mode then the power received by the antenna under test might be 20–40 dB less than this value. Hence the parameters such as transmitted power of the source and the distance between the transmitter and antenna under test is decided by the noise floor of the spectrum analyser and the least power to be received by the antenna under test. As a rule of thumb, if the least measurable gain of the antenna under test is P(dBi) and the noise floor of the spectrum analyser is NL(dBm), then the distance must be adjusted in such a way that P(dBi) is at least 20 dB higher than NL(dBm).

But, there is another catch in deciding the distance between the standard gain horn and the antenna under test. The wavefront being received at the aperture of the test antenna must be planar in nature to avoid errors in measurement, which in mathematical terms means that the phase difference between the ray travelling from path R_1 and the one from R_2 must be close to 0°. An example would increase the insight on this principle. Imagine an antenna under test whose edges of the aperture are separated by 10 mm, $R_1 = 50$ cm and $R_2 = 50.009999$ cm. Now the phase corresponding to R_1 could be calculated as $\varphi_1 = (2\pi R1)/\lambda = 16822.43°$, similarly $\varphi_2 = (2\pi R2)/\lambda = 16825.8°$ translating to a phase difference of 3° which could be concluded as a decent plane wave.

It must also be noted that the indoor measurements of the antenna are also valid, since multipath effects would be minimal at these frequencies.

9.5 Gain Measurements

There are two methods for gain measurement of the antenna under test. The first is using a vector network analyser (VNA), and the second is in an anechoic chamber.

The gain measurement setup with method 1 is depicted in Figure 9.6. The ports of the VNA are connected to the standard gain horns separated by the far-field distance and $|S_{21}|$ is measured across the frequency of interest. Care must be taken to align the antennas

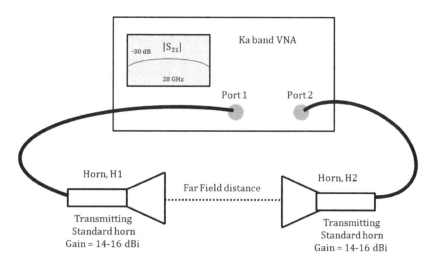

FIGURE 9.6
Setup for gain measurement – Method 1.

for a line of sight link, so that maximum power is received. First, the two horns H1 and H2 are connected to the VNA to measure $|S_{21}|_{HH}$, and Equation 9.1 is used to estimate the gain of the standard horns. Both the horns H1 and H2 must be identical for this measurement. G_H is the measured gain of the standard horn across the frequency of interest. It is must be noted that the VNA would store the S-parameters on a logarithmic scale, which must be converted to a linear scale using $|S_{21}|_{log} = 20\log_{10} S_{21}$. These linear values could be used in Equation (9.1) to compute the value of G_H across the frequency. R is the distance between the aperture of the horn and the aperture of the AUT.

$$|S_{21}|_{HH} = G_H \frac{\lambda}{4\pi R} \tag{9.1}$$

After the standard gain measurement in the setup of Figure 9.6, horn H2 is replaced with the antenna under test. Again for the horn H1 and the antenna under test (AUT), $|S_{21}|$ is measured in the VNA and termed as $|S_{21}|_{HAUT}$. Once, these values are noted, Equation (9.2) could be used to estimate the gain of the AUT termed as G_{AUT}. G_{AUT} could be converted to a logarithmic scale using Equation (9.3).

$$\left(|S_{21}|_{HAUT}\right)^2 = G_{AUT} G_H \left(\frac{\lambda}{4\pi R}\right)^2 \tag{9.2}$$

$$G_{AUT(log)} = 10 \log G_{AUT} \tag{9.3}$$

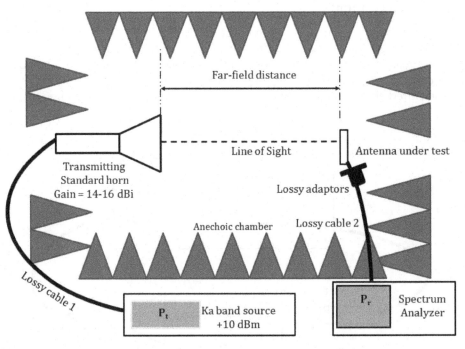

FIGURE 9.7
Setup for gain measurement – Method 2.

For gain measurement using method 2, AUT is placed in an anechoic chamber for received power measurement. The setup for method 2 is shown in Figure 9.7. The cable losses and the adaptor losses must be carefully noted before the computation of gain. The losses in the setup are usually frequency sensitive. Equation (9.4) could be used to estimate the measured gain of the AUT.

P_t^{eff} is the effective transmitted power, which is computed using the following equation, P_t^{eff} (dBm) = P_t (dBm) − Cable loss (dB) − Adaptor loss (dB).

$$P_r = P_t^{eff} G_{AUT} G_H \left(\frac{\lambda}{4\pi R} \right)^2 \tag{9.4}$$

9.6 Conclusion

In this chapter, fabrication methodology for PCB-based antennas was discussed, along with the sources of errors. Tolerance analysis of the variation in the dimensions of the antenna and its effect on the impedance characteristics was also discussed. The requirements for the radiation pattern measurement setup was explained, along with sources of error in the process. S-parameter measurements in the context of mmWaves was also discussed with justification. The VNA and anechoic chamber methods of gain measurement was also explained in detail.

References

1. C. C. Cutler, A. P. King, and W. E. Kock, "Microwave Antenna Measurements," *Proceedings of the IRE*, 35(12), 1462–1471, December 1947.
2. "IEEE Standard Test Procedures for Antennas", ANSI/IEEE Std 149–1979, p. 1, 1979.
3. J. A. G. Akkermans, R. van Dijk, and M. H. A. J. Herben, "Millimeter-wave antenna measurement," *2007 European Microwave Conference*, Munich, 83–86, 2007.
4. A.C.F. Reniers, Q. Liu, M.H.A.J. Herben, and A. B. Smolders, "Review of the accuracy and precision of mm-wave antenna simulations and measurements," *10th European Conference on Antennas and Propagation (EuCAP), Davos, Switzerland*, 1–5, 2016.
5. M. Huang M.H.A.J. Herben A.C.F. Reniers, and P.F.M. Smulders, "Causes of discrepancies between measurements and EM-simulations of millimeter-wave antennas," *IEEE Antennas and Propagation Magazine*, 55(5), 139–149, 2013.
6. P. F. M. Smulders H. Yang, and J. A. G. Akkermans, "On the design of low-cost 60-GHz radios for multigigabit-per-second transmission over short distances," *IEEE Communications Magazine*, 45(12), 44–51, December 2007.
7. L. Chen, A. S. G. Andrae, G. Zou, and J. Liu, Characterization of substrate materials for system-in-a-package applications," *Journal of Electronic Packaging*, 126(2), 195–201, June 2004.
8. Y. P. Zhang and D. Liu, "Antenna-on-chip and antenna-in- package solutions to highly integrated millimeter-wave devices for wireless communications," *IEEE Transactions on Antennas and Propagation*, AP-57(10), 2830–2841, October 2009.

9. D. C. Thompson, O. Tantot, H. Jallageas, G. E. Ponchak, M. M. Tentzeris, and J. Papapolymerou, "Characterization of liquid crystal polymer (LCP) material and transmission lines on LCP substrates from 30 to 110 GHz," *IEEE Transactions on Microwave Theory and Techniques*, 52(4), 1343–1352, April 2004.
10. S. Smith and V. Dyadyuk, "Measurement of the dielectric properties of rogers R/flex 3850 liquid crystalline polymer substrate in V and W band," *IEEE International Symposium on Antennas and Propagation*, 4B, 435–438, July 2005.
11. Y. Lu, Y. Huang, K. Teo, N. Sankara, W. Lee, and B. Pan, "Characterization of dielectric constants and dissipation factors of liquid crystal polymer in 60–80 GHz band," *IEEE Antennas and Propagation Society International Symposium*, San Diego, CA, USA, 1–4, July 2008.
12. M. D. Huang, M. I. Kazim, and M. H. A. J. Herben, "Characterization of the Relative Permittivity and Homogeneity of Liquid Crystal Polymer (LCP) in the 60 GHz band," COST 2100 TD(10)12031, November 2010.

10

Research Avenues in Antenna Designs for 5G and beyond

10.1 Introduction

Even though researchers have been designing antennas at millimeter wave frequencies for decades, it is interesting to note that the designs applicable for cellular communication has been on the rise in recent years. This proves that research targeting antenna designs for 5G mobile terminals and base stations is still in the nascent stage. In this chapter, various research opportunities for antenna designs for 5G and beyond are discussed, with technical insights.

10.2 PCB-Based Antenna Designs for 5G Cellular Devices

Several antenna module designs have been presented in this book, and some research opportunities for the improvement of these designs are listed below:

- The CPW-fed conformal designs could be investigated with localized ground planes and with the integration of absorbers in place of reflectors, leading to much more electrically compact structures. The conformity of the designs presented were a discreet corner bending, a smoother profiled aperture might be designed for state of the art smartphone panels. The length of the conformal designs presented is slightly more than in commercial phone models; if this could be reduced without a compromise in the gain and operational bandwidth, it would be more applicable to industry standard antennas.
- The conformal antennas presented in this book were primarily stand-alone antennas matched to the 50 Ω port. The circuit would be more meaningful if this design could be re-engineered with a power amplifier, low-noise amplifier and a switch to conceive a more realistic design.
- The surface roughness of the reflector presented in the wideband corner bent design could be investigated. This roughness would be important when the frequency of operation is further scaled up, since the corresponding wavelength would reduce and the surface roughness caused by the limiting resolution of the 3D printer might

result in a specular pattern with a degradation in gain. Thus an investigation of the dependence of the surface roughness of the reflector on the integrity of the pattern might be interesting.

- The backward compatibility design of 4G-LTE and mmWave 5G could be expanded for a multi-port scenario. It must also be noted that the proposed 4G design is compatible with only one of the bands of 4G, but this could be enhanced without an increase in the physical footprint. The latest smartphone models have a larger metallic ground which could be investigated for 4G and 5G antennas together. Further antennas catering to commercial standards such as WiFi, Bluetooth etc. might also be explored.

- Even though the overlapped pattern diversity module designed on a polycarbonate substrate was illustrated, the aperture efficiency and impedance bandwidth is relatively low because of the lossy substrate. If an inexpensive substrate with a low dielectric loss tangent could be used to redesign the module it might be relevant to the 5G community. The feeding method incorporating a switch might also be investigated.

- Alternative topologies to achieve orthogonal pattern diversity might be explored, since the physical footprint is slightly bigger than in commercially available phones. An investigation of orthogonal patterns with closer elements and a mechanism for mutual coupling reduction could be novel.

- Aperture efficiency enhancement techniques could be explored in the context of travelling wave antennas centred at 28 GHz. Even though the presented designs have high pattern integrity, impedance bandwidth is limited, and strategies to simultaneously enhance gain and impedance bandwidth without compromise in aperture efficiency might be interesting to investigate.

- Special superstrates might be designed to be integrated with phased array to achieve path loss compensation. Conventionally, phased arrays suffer from scanning loss when the beam is scanned, but if a superstrate or an additional shaped aperture could be designed in such a way that the gain enhancement for the beam corresponding to an angle is in accordance with the path loss compensation equations.

- The sub-wavelength dual-polarized metamaterial unit cell would operate only for orthogonal incident polarizations. If this could be redesigned to accommodate multiple incident polarizations, then a compact beam switching module with a low physical footprint could be realized. A design perspective on integration of mmWave switches would also add value to the proposed design.

- The scanning loss for the corner bent phased array is higher compared to its planar counterpart. This phenomenon could be attributed to higher speckles in the pattern caused by corner bending. If the geometry could be optimized for lower scanning loss, the design could be novel.

- Pattern reconfigurable antennas are typically restricted to lower frequencies. A demonstration of the principle of pattern reconfigurability is depicted in Figure 10.1, along with the corresponding patterns for the individual states of switching. In this design, metallic strips were used as an abstraction for the diodes. If actual diodes with bias lines could be demonstrated, a commercially viable solution might be realized.

Research Avenues in Antenna Designs for 5G and beyond 215

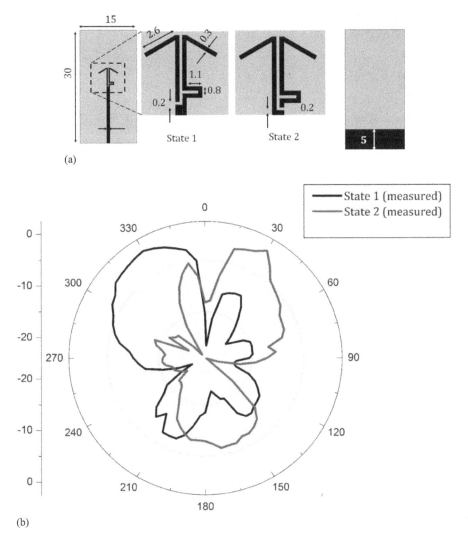

FIGURE 10.1
(a) Reconfigurable angled dipole operating at 28 GHz and (b) End-fire patterns for both the states [19].

10.3 Application of Additive Manufacturing for Antennas

Additive manufacturing is a mechanical process of fabrication where material deposition happens according to the desired geometry [1–5]. This is in stark contrast with subtractive manufacturing, in which material is removed depending on the geometry. 3D printing is a famous example of additive manufacturing. In this process, a known dielectric filament roll is heated to a precise temperature and deposited according to the solid geometry.

FIGURE 10.2
Source antennas and their 3D-printed lenses [1].

The user has to design the geometry in any computer aided design (CAD) software and this file would be used as an input in commercial 3D printers. The important feature of 3D printing is that the printer itself is inexpensive and the time needed for printing is relatively low, indicating a high throughput.

3D printing is limited by the resolution of the printer. In other words, the resolution of the printer decides the least dimension that could be printed without error with that printer. The lower the resolution, the better for the antennas operating in the Ka band and beyond. The electrical characteristics of the material used in 3D printing must first be characterized for a better understanding of the material's behaviour before it can be used for antenna design. The dielectric constant and dielectric loss tangent of the dielectric to be used in 3D printing must be verified experimentally prior to the antenna design.

Figure 10.2 depicts an integrated lens antenna fabricated using fused deposition modelling (FDM) with polylactic acid (PLA) as the dielectric. Here, the lens is manufactured using 3D printing but the source antenna is fabricated using conventional methods. The antenna integrated with the lens operates in the 28 GHz band with 10.7% impedance bandwidth yielding a gain of 15.6 dBi. The radius of the lens is 16 mm, translating to almost 1.6λ at the frequency of operation.

With reference to 3D printing, its useful features must be exploited to realize a reasonably good design. One of the possibilities is that conventional substrates used for designing PCB-based antennas suffer from dielectric loss, which would hamper the antenna characteristics, especially in the mmWave band. If the substrate could be carved out selectively with 3D printing then radiation efficiency, bandwidth and gain could be increased.

Most of the reported 3D printed lens designs are operational only with boresight radiation. If gain equalized beam scanning with a 3D printed lens could be produced, it could be novel. The lens behaviour for beam collimation must be such that, irrespective of the angle of incidence from any source antenna, the gain yield must be same. This might be achieved by sculpting the 3D profile of the lens with respect to the angle of incidence. A design example is presented in the following section.

10.3.1 A Dual Band mmWave Antenna on 3D Printed Substrate

The schematics of the proposed insert-fed dual band mmWave antenna is illustrated in Figure 10.3. It is designed using polylactic acid (PLA), with dimensions of 20 mm × 20 mm

Research Avenues in Antenna Designs for 5G and beyond

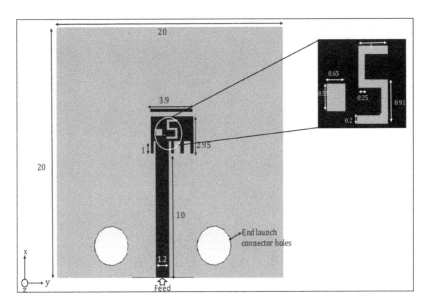

FIGURE 10.3
Proposed planar antenna (all dimensions in mm).

with relative dielectric constant (ε_r) of 2.75, loss tangent (tan δ) of 0.01, and thickness of substrate as 0.5 mm. Copper of thickness 0.017 mm is used as the conductive material. The insert feed line is designed with a width of 1.2 mm according to equations below,

The gap between the patch and the feed line is 0.2 mm, which, translated to an impedance of 52 Ω, lead to the proper matching of the antenna at the feed point. The inset feed line width was chosen to accommodate the south-west connector. Design of the antenna can be described in two steps. First, a microstrip patch antenna with insert feed is designed with operating frequency of 28 GHz. Next, slots are etched out of the patch according to detailed parametric analysis, such that we get 38 GHz as another resonating frequency. First, the S-shaped slot is etched out, and the operating frequencies obtained are 28 GHz, 36 GHz and 38 GHZ. To eliminate 36 GHz as one of the operating frequencies, another square-shaped slot is etched on to the patch. After obtaining the desired frequencies, a parasitic element is added to improve the gain of the proposed antenna. After designing the proposed antenna, corners of PLA substrate are eliminated to obtain the same results. This shows that antenna can be practically fabricated on PLA, which offers the best choice for 3D printing. Figure 10.4(a) shows the two stages of the antenna's development, and Figure 10.4(b) shows the same circuit with chopped-off substrate. Dimensions of the chopped portion are 6 mm from the port and approximately λ/2 (i.e. 5 mm) lengthwise.

The most important part is to 3D print the proposed antenna. We use the chemical etching method to fabricate the antenna design. In this method we basically prepare a mask of the desired antenna and then a strip of copper is pasted on to the substrate. Then the remaining part (other than the masked portion of the copper strip) is milled out to make a proper antenna. Figure 10.4(c) shows the fabricated antenna.

The performance of the proposed antenna is analysed and optimized using CST Studio Suite 2018. The simulated and measured input reflection coefficients of the proposed 3D printed dual band antenna for mmWave 5G applications are depicted in Figure 10.5. It is obvious that the antenna is matched at both the operating frequencies with $|S_{11}|$ less than −10 dB. It is observed that the bandwidth is 6% at centre frequency 29 GHz, and 5.3% at

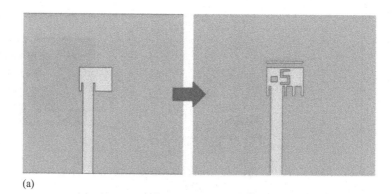

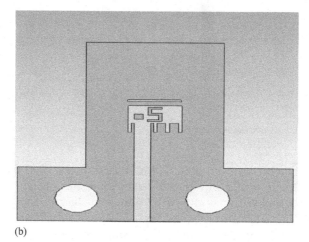

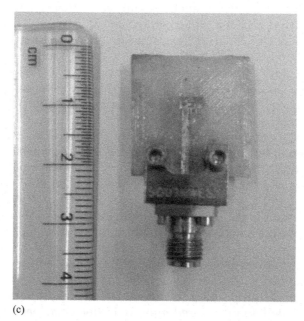

FIGURE 10.4
(a) Stages of antenna design, (b) Antenna with shaped substrate and (c) Photograph of the fabricated prototype.

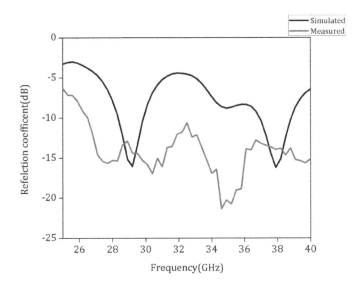

FIGURE 10.5
Simulated and measured return loss of the proposed antenna.

centre frequency 38 GHz. Slight differences in simulated and measured values are caused by the crude method of fabrication.

Radiation patterns of the 3D printed antenna in E-plane operating at frequencies 28 GHz and 38 GHz are depicted in Figure 10.6(a) and 10.6(b), respectively. The difference between simulated and measured results is because of the high loss tangent of the substrate used, and the connectors used in measurements.

Figure 10.7 shows a gain versus frequency plot. As observed, the antenna radiates with high gain at both the operating frequencies, making it applicable for 5G communication.

10.4 On-chip Antennas for CMOS Circuitry

These are the antennas compliant with the industry standard CMOS fabrication process [6–11]. Most of the transceiver circuitry operating in the 28 GHz band could be readily designed using CMOS process flow. In this context, PCB-based antennas designed on conventional substrates such as Rogers or Nelco would require additional matching circuits and interconnects that connect the radio frequency integrated circuit (RFIC) to the antenna. This strategy would increase the overall footprint of the circuit. This technique would also increase the insertion loss caused by the additional impedance transformers connecting the bond wires to the antenna, leading to gain degradation of the integrated antenna.

Therefore, on-chip antennas are a promising solution to this problem. Typically, these must be designed on a lossy silicon substrate. Figure 10.8 shows the layers in a typical CMOS fabrication which has been realized with 28 nm CMOS process flow. The micrograph of the on-chip antenna operating at 33 GHz integrated with the RFIC is also depicted in Figure 10.8.

The radiation efficiency of on-chip antennas is typically in the range 20%–30% as it is designed on a high dielectric constant silicon substrate. This is one of the primary

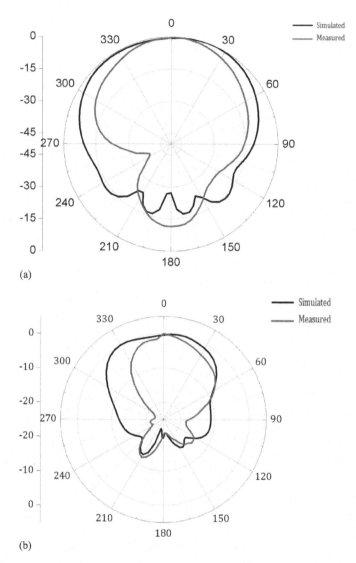

FIGURE 10.6
Broadside radiation patterns at (a) 28 GHz and (b) 38 GHz.

challenges in on-chip antenna design. Designs that are compliant with the CMOS process flow with high radiation efficiency, and high gain is the need of the hour.

Another interesting design aspect is the packaging of the chip. RFIC packaging is typically lossy, but if this could be redesigned for gain enhancement of the underlying on-chip antenna, this design approach could compensate for gain degradation caused by the lossy substrate. A design example is presented below.

10.4.1 A Wideband CPS-Fed Dipole on Silicon

The coupled transmission lines have been recreated in a circuit simulator (ADS) and in a full wave simulator (HFSS). The 90° lines have a gap between them of 50 μm, and the width

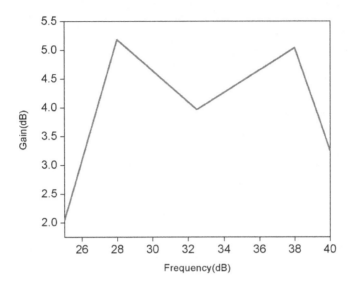

FIGURE 10.7
Broadside gain of the antenna.

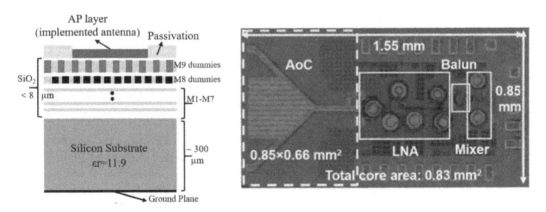

FIGURE 10.8
Layers in a CMOS circuit along with the fabricated chip [6].

of the lines is 30 μm each, designed on a 5 mil Rogers RO3003 substrate. The response of the coupled lines alone is shown in Figure 10.8(a). It can be observed that the quarter-wave coupled microstrip transmission lines behave as a bandpass filter from 24 to 36 GHz, the slight deviations between the curves possibly being attributed to the poor abstraction of microstrip discontinuities in the circuit simulator. The dual of the same network was also simulated in ADS and HFSS, where a very sharp band notch at 28 GHz, which is in congruence with the proposed design with $Z_{oe} = 175\ \Omega$ and $Z_{oo} = 87\ \Omega$, is observed. The dimensions are suitable for the implementation of on-chip passive components in CMOS, specifically in the frequencies above 20 GHz.

The impedance bandwidth of a conventional branch line coupler has been increased to 40%. A similar design strategy would be incorporated for feeding an antenna, which would include the popular 5G candidate, 28 GHz band. Coupled line transmission lines are used as an impedance matching network between the CPW feed and the radiating arms of the antenna.

Wideband antenna topologies include monopole-based designs, and aperture-coupled, slot-loading and probe-fed designs. Even though printed monopole designs could give more than 100% impedance bandwidth, the gain at the higher frequencies would be low. Aperture coupled antennas would give more than 60% bandwidth, but implementation on silicon would pose fabrication challenges, because of the multi-layer design. Slot-loading would also help in impedance bandwidth enhancement but the pattern integrity for wide bandwidth is questionable. Design modifications to probe-fed antennas would also increase cross-polarization levels. Apart from higher feed loss, these effects would be dominant at frequencies above 25 GHz because of the comparable size of the probe with respect to the quarter wavelength.

Uniplanar designs are preferred for integration with conventional bulk CMOS processes, hence CPW-fed topology is chosen for the proposed antenna. Impedance match would be difficult to realize at 28 GHz on a thin substrate (20 mil) such as Rogers 5880 ($\varepsilon_r = 2.2$), since the gap and widths would be 10 µm and 25 µm, respectively. But these dimensions would easily be possible using the standard silicon process. A CPW-fed antenna proposed in previously published articles has a narrow bandwidth because of strongly resonant structures. CPW to slot-line baluns would improve the bandwidth, but gain would be frequency selective. Thus CPW feed integrated with coupled transmission lines is presented in this section to achieve an impedance bandwidth over 40% with stable patterns across the band.

Typical on-chip antennas are suitable for frequencies beyond 60 GHz, as the dimensions are readily compliant with this frequency of operation. Since the dimensions of the silicon die match the radiator, these frequencies are popular. IBM's Thomas J. Watson Research Center has reported a complete on-chip solution to phased array at 28 GHz, with the antenna wide-beam and narrowband. A leaky-wave structure with high gain and beam steering capability would be unsuitable for deployment since the frequency would be fixed a priori. The proposed design is 1 µm aluminium etched out on 1 µm SiO_2 layer ($\varepsilon_r = 4$), which is grown on 275 µm bulk silicon ($\varepsilon_r = 11.9$). A silicon dioxide layer is grown to improve the radiation efficiency by reducing the effective dielectric constant. The coupled lines match the impedance of the dipoles and the CPW feed, as shown in Figure 10.9. All the dimensions are in mm. The primary radiators are of length 1.9 mm, optimized for wideband. The distance between CPW ground and radiators is optimized for minimal mutual coupling and an overall smaller electrical footprint.

The CPW feed is designed to match the input impedance of the dipole arms on the silicon substrate. It must be noted that the gap between the radiating arms is optimized for minimal beam-tilt caused by asymmetric feed. The gap between the coupled lines in the proposed design is 50 µm, which translates to 0.5% of the wavelength at the highest frequency of operation, thus stable end-fire patterns are obtained throughout the band. With an increased gap between the coupled lines, the difference between even and odd mode impedances would be high, hence creating a mismatch with the radiator's input impedance. A smaller gap would enhance the impedance bandwidth but might create fabrication difficulties.

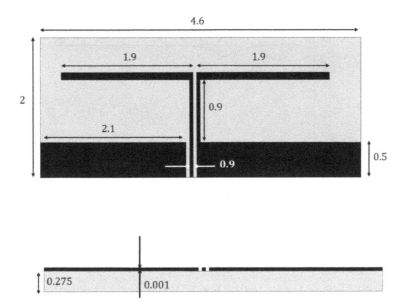

FIGURE 10.9
Top and side view of the proposed on-chip silicon antenna

The line width is also optimized for impedance match, as observed in Figure 10.10(a); a thinner line with a wider gap would create a stronger mismatch. Figure 10.10(a) also depicts the return loss for the variation of the radiator's length; the smaller the length, the higher the frequency of operation. The optimized length of each dipole arm for wideband operation is 1.9 mm (0.12λ at 19 GHz and 0.19λ at 30 GHz), the width of the radiator is primarily the width of the coupled lines, and thinner lines are preferred at frequencies beyond 60 GHz. The overall size of the antenna was also optimized for compact size at 19 GHz (0.29λ × 0.12λ). Further miniaturization would create a narrow band, hence the proposed dimensions are a trade-off between impedance bandwidth, gain and electrical footprint. The bandwidth of the proposed design is 10.4 GHz (19–29.4 GHz, 43%), as illustrated in Figure 10.10(b). Here, the return loss is up to −25 dB, indicating high radiation efficiency. Tapered feed lines were also investigated, but the impedance bandwidth of coupled lines was higher than the tapered lines. The impedance bandwidth could be further enhanced by the incorporation of spirals or split-ring resonators as radiators, which would result in a frequency selective gain.

The normalized radiation patterns are shown in Figure 10.11 at 20, 24 and 28 GHz, respectively. The H-plane patterns appear omnidirectional, because of the lack of metallic ground in the feed; this is an inherent feature of CPW-fed designs. The E-plane pattern experiences a minimal beam tilt of 6° at 19 GHz and 2° at 29 GHz. This is attributed to the gap between the coupled lines, thus feeding the antenna asymmetrically. The front-to-back ratio is 2.3 dB across the band, which could be improved by increasing the ground plane or by an end-fire gain enhancement technique such as parasitic patch loading or metamaterial loading. The end-fire gain is 3 dBi for entire the band, as shown in Figure 10.12.

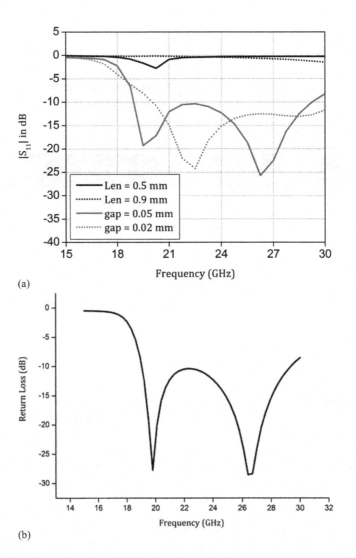

FIGURE 10.10
(a) Return loss for various parameters and (b) Return loss of the proposed antenna.

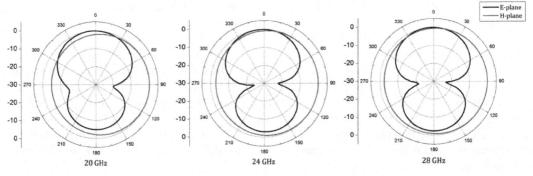

FIGURE 10.11
Radiation patterns at 20, 24 and 28 GHz.

Research Avenues in Antenna Designs for 5G and beyond

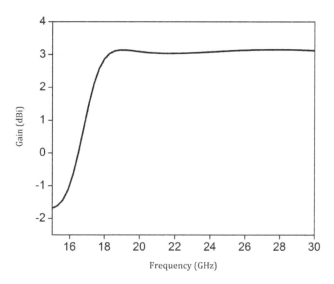

FIGURE 10.12
End-fire gain of the proposed antenna.

10.5 Optically Transparent Antennas

5G researchers state that 5G would also include vehicular data connectivity. In this regard, antennas installed on windscreens would serve the purpose. Hence optically transparent antennas is an important domain of research.

Optically transparent antennas require high optical transparency without deteriorating the conventional characteristics of the radiator, such as impedance bandwidth and gain [12–18]. To be optically transparent, the antenna must have the smallest metallic footprint. But to have high gain the antenna must include a high metallic footprint. Thus it is a contradictory requirement to simultaneously achieve optical transparency and high gain in the mmWave region.

As a compromise, researchers use highly conductive and electrically thin silver-based mesh geometry to realize the antenna, as demonstrated in Figure 10.13, which is a 60 GHz

FIGURE 10.13
Optically transparent antenna array [12].

antenna array with high optical transparency. The silver trace has high conductivity, which aids operations in the mmWave region. The mesh pattern of the antenna would aid in optical transparency. The challenge to achieve high gain and optical transparency is open for research.

10.6 Conclusion

In this chapter, research avenues for designs targeting mmWave 5G frequency band antennas have been explored exhaustively. The research opportunities for designs based on PCB antennas have been explained with specific technical insights. The research gaps of the previously presented designs have been explored. The application of the additive manufacturing process for mmWave antenna designs has been explained with insights for novel designs. Research gaps in the domain of on-chip and optically transparent antennas has been discussed.

References

1. B. T. Malik, V. Doychinov, S. A. R. Zaidi, I. D. Robertson, and N. Somjit, "Antenna gain enhancement by using low-infill 3D-printed dielectric lens antennas," *IEEE Access*, 7, 102467–102476, 2019.
2. V. M. Pepino, A. F. da Mota, A. Martins, and B.-H. V. Borges, "3-D-printed dielectric metasurfaces for antenna gain improvement in the Ka-band," *IEEE Antennas Wireless Propagation Letters*, 17(11), 2133–2136, November 2018.
3. R. Sauleau, C. A. Fernandes, and J. R. Costa, "Review of lens antenna design and technologies for mm-wave shaped-beam applications," *Proceedings of the 11th International Symposium on Antenna Technology and Applied Electromagnetics (ANTEM)*, St Malo, France, 1–5, 2005.
4. K. X. Wang and H. Wong, "A wideband millimeter-wave circularly polarized antenna with 3-D printed polarizer," *IEEE Transactions on Antennas Propagation*, 65(3), 1038–1046, March 2017.
5. R. Gonalves, P. Pinho, and N. B. Carvalho, "3D printed lens antenna for wireless power transfer at Ku-band," *Proceedings of the 11th European Conference on Antennas and Propagation (EUCAP)*, Paris, 773–775, March 2017.
6. M. K. Hedayati et al., "Challenges in on-chip antenna design and integration with RF receiver front-end circuitry in nanoscale CMOS for 5G communication systems," *IEEE Access*, 7, 43190–43204, 2019.
7. M. Wojnowski, C. Wagner, R. Lachner, J. Böck, G. Sommer, and K. Pressel, "A 77-GHz SiGe single-chip four-channel transceiver module with integrated antennas in embedded wafer-level BGA package," *Proceedings of the IEEE 62nd Electronic Components and Technology Conference*, San Diego, CA, USA, 1027–1032, May 2012.
8. C. C. Liu and R. G. Rojas, "V-band integrated on-chip antenna implemented with a partially reflective surface in standard 0.13μm BiCMOS technology," *IEEE Transactions on Antennas Propagation*, 64(12), 5102–5109, December 2016.
9. D. Gang, H. Ming-Yang, and Y. Yin-Tang, "Wideband 60-GHz on-chip triangular monopole antenna in CMOS technology," *Proceedings of the 3rd Asia–Pacific Conference on Antennas and Propagation (APCAP)*, Harbin, P.R.China, 623–626, July 2014.
10. F. Gutierrez, K. Parrish, and T. S. Rappaport, "On-chip integrated antenna structures in CMOS for 60 GHz WPAN systems," *Proceedings of the 3rd IEEE Global Telecommunication Conference (GLOBECOM)*, Honolulu, HI, USA, 1–7, November 2009.

11. M. K. Hedayati, A. Abdipour, R. S. Shirazi, M. John, M. J. Ammann, and R. B. Staszewski, "A 38 GHz on-chip antenna in 28-nm CMOS using artificial magnetic conductor for 5G wireless systems," *Proceedings of the IEEE 4th International Conference on Millimetre-Wave and Terahertz Technologies (MMWaTT)*, Tehran, Iran, 29–32, December 2016.
12. A. Martin, O. Lafond, M. Himdi, and X. Castel, "Improvement of 60 GHz transparent patch antenna array performance through specific double-sided micrometric mesh metal technology," *IEEE Access*, 7, 2256–2262, 2019.
13. T. Yasin, R. Baktur, and C. Furse, "A study on the efficiency of transparent patch antennas designed from conductive oxide films," *Proceedings of the IEEE International Symposium on Antennas and Propagation (APSURSI)*, Spokane, WA, USA, 3085–3087, July 2011.
14. H. J. Song, T. Y. Hsu, D. F. Sievenpiper, H. P. Hsu, J. Schaffner, and E. Yasan, "A method for improving the efficiency of transparent film antennas," *IEEE Antennas Wireless Propagation Letters*, 7, 753–756, 2008.
15. J. Hautcoeur, F. Colombel, X. Castel, M. Himdi, and E. M. Cruz, "Optically transparent monopole antenna with high radiation efficiency manufactured with silver grid layer (AgGL)," *Electronic Letters*, 45(20), 1014–1016, September 2009.
16. M. A. Malek, S. Hakimi, S. K. Abdul Rahim and A. K. Evizal, "Dual-band CPW-fed transparent antenna for active RFID tags," *IEEE Antennas and Wireless Propagation Letters*, 14, 919-922, 2015.
17. S. Y. Lee, D. Choi, Y. Youn and W. Hong, "Electrical characterization of highly efficient, optically transparent nanometers-thick unit cells for antenna-on-display applications," *2018 IEEE/MTT-S International Microwave Symposium, IMS*, IMS, Philadelphia, PA, USA, 2018, 1043–1045.
18. Desai, A, Upadhyaya, T, Palandoken, M, Gocen, C. "Dual band transparent antenna for wireless MIMO system applications," *Microw Opt Technol Lett*, 61, 1845–1856, 2019.
19. G. S. Karthikeya, M. P. Abegaonkar, and S. K. Koul, "Gain enhanced mmWave pattern reconfigurable Quasi angled dipole with minimal DNG loading," *2018 IEEE Indian Conference on Antennas and Propogation (InCAP)*, Hyderabad, India, 1–4, 2018.

Appendices

Appendix A: Hints for Simulations in Ansys HFSS

Ansys HFSS is industry standard software that is very popular for full-wave electromagnetic simulations. Simulations for lower frequency antennas below 10 GHz are straightforward, with no need for special consideration. But when designing mmWave antennas beyond 25 GHz, several aspects of simulations have to be examined. This section deals with hints for the simulation of Ka band antennas for 5G applications.

A.1 Modelling

Care must be taken to use only lossy dielectrics and the actual conductor to be used. For example, copper of finite thickness is recommended rather than a perfect electric conductor (PEC) with no thickness. The dielectric loss tangent and conductor losses would lead to deterioration in the effective gain of the antenna [1–3]. The end-launch connector might be included in the modelling provided the computational resources for simulations support this. In most of the cases, a 50 Ω port impedance might suffice. The uncontrolled currents of the connector might influence the measurement results. The size of the radiation box must be fixed with respect to the lowest frequency of consideration for the sweep.

Figure A.1 illustrates the dialogue box for the solution setup for the Ka band antenna. The user must know a priori about the expected frequency of operation. If the structure is resonant, then a single frequency solution could be used. In the case of broadband structures, a corresponding radio button could be used. The solution frequency always has to be set as the highest frequency of sweep the user would be employing for simulations. For example, if the antenna is expected to operate from 25 to 40 GHz, the solution frequency must be set as 40 GHz. This strategy is to maintain the smallest size of the tetrahedron so that fine results could be achieved. The lower frequency setting in the solution frequency might result in a faulty shift in the resonant frequency. Also, a lower value is selected for solution frequency, so the effects of finer corners and corrugations might be totally missed by the simulation engine. The number of passes must be set at around 20–30, depending on the complexity of the geometry in simulation and the convergence criterion. The S-parameter error value could be set as 0.02; a finer value of this would give a better result but at the cost of greater time and computational resources.

Figure A.2 illustrates the dialogue box of the validation check before simulation. The designer must verify that all the entities are validated. The software would also flag up issues in the warning and errors window. Post-simulation, the designer must verify that the convergence criterion has been met. A typical example is shown in Figure A.3; the target criterion was 0.02 and this was achieved, indicating minimal distortion in the simulated results.

Analysis of the far field radiation patterns with respect to the antenna is an important aspect of simulations. Figure A.4 demonstrates the E-fields for a typical broadside radiator.

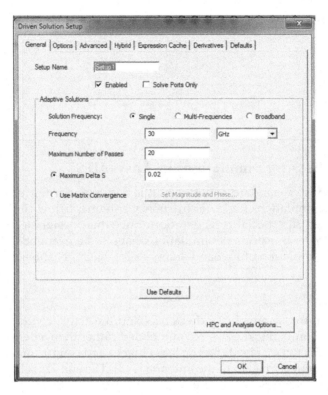

FIGURE A.1
Solution setup dialogue box for a 5G antenna in HFSS.

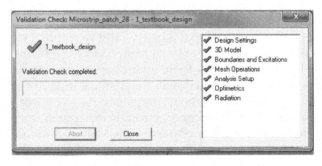

FIGURE A.2
Model validation dialogue box in Ansys HFSS.

The designer must anticipate the source of radiation by analysing the field lines of the entire antenna structure and the corresponding co-polarized and cross-polarized radiation patterns in the principal planes. It is critical to understand that all the field lines do not necessarily contribute to radiation. In this case, the patch antenna is an aperture radiator, wherein the radiation happens from the radiating edges, as shown by the solid lines near

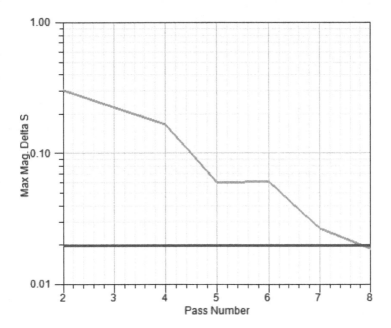

FIGURE A.3
Convergence curve from Ansys HFSS.

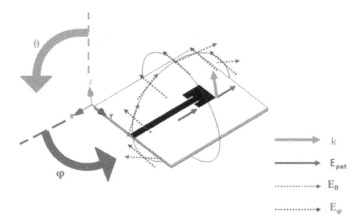

FIGURE A.4
Anatomy of E-fields for a broadside radiator for E-plane.

the inset of the patch. The field responsible for dominant co-polarized radiation is from E_{pat} and the direction of wave propagation is along the Z-axis as indicated by k. Hence the pattern would lead to broadside radiation along the Z-axis. The principal cuts for this radiation are the XZ plane, which is scanning the E-plane, and the YZ plane, which is scanning the H-plane, as demonstrated in Figure A.5. The corresponding unit vectors for the principal cuts is also shown in the respective figures.

A tapered slot antenna is a typical end-fire radiator and the corresponding analysis of the radiation and its far-field is demonstrated in Figures A.6 and A.7.

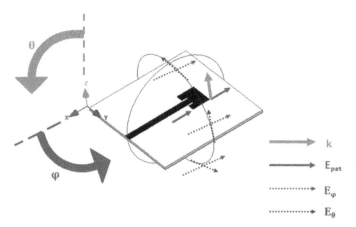

FIGURE A.5
Anatomy of E-fields for a broadside radiator for H-plane.

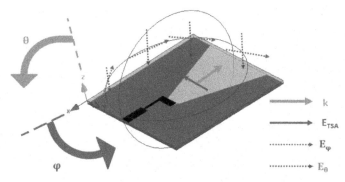

FIGURE A.6
Anatomy of E-fields for an end-fire radiator for E-plane.

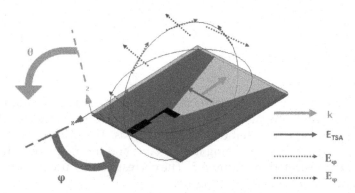

FIGURE A.7
Anatomy of E-fields for an end-fire radiator for H-plane.

Appendix B: Measurement Issues with End-Launch Connector

The connector assembly with the antenna or device under test is critical for measurements, especially in the Ka band [4–5]. Conventional commercial SMA connectors would operate up to 26.5 GHz. This does not mean that it would fail beyond the rated frequency. If the SMA connector could be soldered with a precision thin-tip soldering iron, then the possibility of it working would be higher. The trace of the SMA connector is close to 1 mm but most of the feed lines designed on electrically thin substrates have feed line widths much less than this, thus making assembly difficult. The mismatch between the traces might create additional distortion in the S-parameter's measurement.

Hence end-launch connectors rated at 40 GHz are an ideal choice for the measurement of mmWave 5G antennas. The designer must allow ample clearance for the connector holes as per the official manual of the connector. Figure B.1 shows the exploded view of an end-launch connector. The primary entity has the trace pin and the upper ground mould. This trace pin must coincide with the feed line of the fabricated antenna. It is recommended that the trace pin must be soldered to the device under test. More often than not, soldering might not be necessary, as demonstrated in various chapters of this book. The grounding plates must be assembled precisely and fastened with an appropriate Allen key. An example of the assembled end-launch connector is shown in Figure B.2.

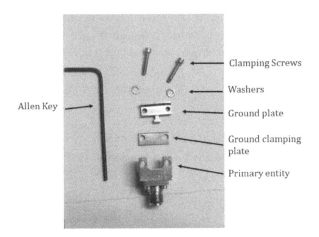

FIGURE B.1
Exploded view of part of an end-launch connector.

FIGURE B.2
Isometric view of an assembled end-launch connector.

Appendix C: Material Parameters' Extraction Using S-parameters

The S-parameters simulation for a unit cell was explained in section 5.2.3. In this section, material parameters' extraction using MATLAB is illustrated [6–8]. Initially, a Floquet port based simulation of the unit cell must be performed in HFSS. The data of real and imaginary components for the S-parameters of the unit cell simulation must be imported into the MATLAB workspace. Then the following code could be used for the extraction of material parameters.

```
S11=reS11+(i*imS11);   %imported from the simulations of unit cell
S21=reS21+(i*imS21);   %imported from the simulations of unit cell
d=5e-3;                % the primary distance between ports in HFSS
f=FreqGHz*1e9;         % convert to actual units, since HFSS exports in
the GHz scale
lambda=(3e8)./f;       %calculate wavelength

% routine for calculation of mu and epsilon
% j is chosen to avoid the imaginary notation 'i'
% 51 discrete points were assumed in the simulation
for j=1:51
       K(j)=((S11(j)^2)-(S21(j)^2)+1)/(2*S11(j));
       Gamma(j)=K(j)-sqrt((K(j)^2)-1);
       T(j)=(S11(j)+S21(j)-Gamma(j))/(1-((S11(j)+S21(j))*Gamma(j)));
       Gumma(j)=(log(1/T(j)))/d;
       Gumma0(j)=i*((2*pi)/lambda(j));
       epsilon(j)=(Gumma(j)*(1-Gamma(j)))/(Gumma0(j)*(1+Gamma(j)));
       epsi(j)=conj(epsilon(j));
       mu(j)=(Gumma(j)*(1+Gamma(j)))/(Gumma0(j)*(1-Gamma(j)));
       mu_final(j)=conj(mu(j));
end
```

```matlab
eps_Real=real(epsi);      % real value of epsilon
 figure(1);
 plot(FreqGHz,eps_Real);
 title('eps Real');

 figure(2);
eps_im=imag(epsi);        % imaginary value of epsilon
plot(FreqGHz,eps_im)
title ('eps imaginary');

 figure(3);
mu_real=real(mu_final); % real value of mu
plot(FreqGHz,mu_real);
title('mu_real');

figure(4);
mu_ima=imag(mu_final);
plot(FreqGHz,mu_ima);

figure(5);
plot(FreqGHz,eps_Real,'color','r'); hold on;
plot(FreqGHz,eps_im,'color','b');
title('permitivitty');

figure(6);
grid on;
plot(FreqGHz,mu_real,'color','r'); hold on;
plot(FreqGHz,mu_ima,'color','b');
title('permeability');
legend('real','imaginary');

figure(7);
grid on;
plot(FreqGHz,eps_Real,'color','r'); hold on;
plot(FreqGHz,eps_im,'color','b');
title('permitivitty');
legend('real','imaginary');
```

Appendix D: Useful MATLAB Codes

Aperture efficiency: to calculate the aperture efficiency of an antenna, it is important to initially understand the physical aperture of the antenna and then contemplate the actual gain [9,10]. Figure D.1 illustrates a broadside radiator integrated with a superstrate of

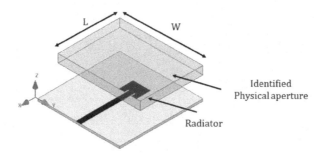

FIGURE D.1
Aperture efficiency for a broadside radiator.

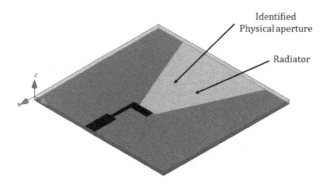

FIGURE D.2
Aperture efficiency for an end-fire radiator.

dimensions $L \times W$. The effective physical aperture of this topology could be identified as the physical area of the superstrate, hence A_p is the area of the superstrate, i.e., LW. From this entity the gain feasible from the physical aperture G_p could be calculated using Equation D.1, where λ is the wavelength in free space. The actual gain, G_{act} is the value from measurement or simulations. The ratio of actual gain to physical gain gives the aperture efficiency. It is imperative to divide the gain values in linear scale as against the logarithm scale, which leads to faulty high ratios.

Figure D.2 The scenario for a typical end-fire aperture ratio. The value of the physical aperture could be calculated using first principles or from the software.

$$G_p = (4\pi A_p)/\lambda^2 \tag{D.1}$$

The following MATLAB code is useful for extracting aperture efficiency.

```
f=FreqGHz*1e9;         % convert to actual units
lambda=(3e8)./f;       % calculate wavelength
Ap=180*1e-6; %in m^2   % Physical aperture either calculated or from HFSS
                       % j is chosen to avoid the imaginary nottaion 'i'
```

```
for j=1:21
    Gp(j)=(4*pi*Ap)/(lambda(j).^2);   % G_p calculation
    Ge(j)=Gain(j);                    % G_act (in linear scale) from HFSS
    Ae(j)=Ge(j)./Gp(j);               % Aperture efficiency as ratio
    Ae_percent(j)=Ae(j).*100;         % Aperture efficiency as percentage
end
```

The Envelope Correlation Coefficient (ECC) is an important metric when designing closely spaced antenna elements. The input reflection coefficient and the mutual coupling is extracted from HFSS. The following code could be used for ECC calculation.

```
S11=reS11+i*imS11;              % S_11 imported from HFSS
S11_star=conj(S11);
S21=reS21+i*imS21;              % S_21 imported from HFSS
S21_star=conj(S21);
modS11=magS11;
modS21=magS21;
S11_double=modS11.^2;
S21_double=modS21.^2;

for j = 1:51
    alpa(j)=(S11_star(j).*S21(j))+(S21_star(j).*S11(j));
    alpa2(j)=alpa(j).^2;
    deno(j)=(1-S11_double(j)-S21_double(j));
    ECC(j)=alpa2(j)./deno(j);
end
```

References

1. https://www.ansys.com/products/electronics/ansys-hfss.
2. https://www.ansys.com/-/media/ansys/corporate/resourcelibrary/techbrief/ab-ansys-hfss-for-antenna-simulation.pdf?la=ja-jp&hash=5DDAE142D173937E4C4946C0AB877EA1A856B401.
3. https://www.ansys.com/services/training-center/electronics/ansys-hfss-getting-started.
4. https://mpd.southwestmicrowave.com/product-category/end-launch-connectors/.
5. https://mpd.southwestmicrowave.com/wp-content/uploads/2018/07/Optimizing-Test-Boards-for-50-GHz-End-Launch-Connectors.pdf.
6. A. B. Numan and M. S. Sharawi, "Extraction of material parameters for metamaterials using a full-wave simulator [Education Column]," *IEEE Antennas and Propagation Magazine*, 55(5), 202–211, October 2013.
7. S. Arslanagic, T. V. Hansen, N. A. Mortensen, A. H. Gregersen, O. Sigmund, R. W. Ziolkowski, and O. Breinbjerg, "A review of the scattering-parameter extraction method with clarification of ambiguity issues in relation to metamaterial homogenization," *IEEE Antennas and Propagation Magazine*, 55(2), 91–106, April 2013.

8. D. R. Smith, S. Schultz, P. Markos, and C. M. Soukoulis, "Determination of effective permittivity and permeability of metamaterials from reflection and transmission coefficients," *Physics Review B*, 65(19), 195104, April 2002.
9. T. H. Jang, H. Y. Kim, I. S. Song, C. J. Lee, J. H. Lee, and C. S. Park, "A wideband aperture efficient 60-GHz series-fed E-shaped patch antenna array with copolarized parasitic patches," *IEEE Transactions on Antennas and Propagation*, 64(12), 5518–5521, December 2016.
10. Y. He, N. Ding, L. Zhang, W. Zhang, and B. Du, "Short-length and high-aperture-efficiency horn antenna using low-loss bulk anisotropic metamaterial," *IEEE Antennas and Wireless Propagation Letters*, 14, 1642–1645, 2015.

Index

A

absorber, 33
additive manufacturing, 215
alignment error, 29
anechoic chamber, 29
angular coverage, 9, 22, 130, 133, 137, 140, 189
Ansys HFSS, 28, 46, 59, 77, 94, 204, 229
antenna under test, 37, 208–210
aperture, 7–8, 12, 22, 26, 28, 30, 32–34, 37–40, 45, 4–50, 60, 65–66, 74, 76–86, 122, 124, 126, 138, 139, 140, 149–160, 181, 200, 204, 208–214, 222, 230
aperture efficiency, 100, 102, 106, 117, 119, 121, 125, 127–128, 133, 136, 139–142, 143, 145, 214, 235–237
architecture, 2, 3, 6, 12
attenuation, 3–4, 7, 11, 24, 92, 121, 192

B

back lobe, 33, 79, 82, 180, 209
base station, 6, 115, 149
beam scanning, 9, 12, 22, 49, 64, 189–192, 196, 199–200, 216
beam switching, 53, 130, 158, 172, 189, 214
beam tilt, 9, 12, 82, 86, 115–116, 151, 170, 223
beamwidth, 8, 22, 32
bending, 21, 26, 30, 32, 45, 59, 70, 79–80, 85, 92, 94, 98, 99, 176, 181, 194, 197, 213–214
bondwire, 57
boresight, 12, 24, 55, 116, 120, 138, 149, 151, 157, 172, 190, 192–193, 196–197, 199, 216
broadside, 11–12, 23–24, 26, 29–30, 41, 55, 76, 91, 119, 131–132, 191–192, 197, 220, 229, 231, 235

C

cellular, 1–4, 6–9, 11, 92, 115
cellular layout, 6
characteristic impedance, 41, 44, 66, 76, 85, 94–95, 167
chemical etching, 24, 55, 57, 66, 136, 204, 217
chemical vapour deposited, (CVD) 58
co-design, 5, 11–12, 14, 164, 181, 184
complementary metal oxide semiconductor (CMOS), 12, 189, 219–222
composite right/left handed (CRLH), 164–166, 168, 172, 185
computer aided design (CAD), 216
conformal, 13, 14, 16, 21, 23–49, 74, 79–111, 145, 163–172, 185, 213
controllers, 11, 64, 149, 193
coplanar stripline (CPS), 12, 41, 76–77, 91–92, 220
coplanar waveguide (CPW), 11, 13, 21, 23–49, 73–77, 85, 89, 91–92, 116, 167–168, 222–223
co-pol, 26, 32, 37, 95, 115, 137, 140, 168, 208, 230–231
copper film, 21, 45–46, 57–58
corner bent, 23–24, 30–50, 56, 60, 176, 181, 189–200
coverage, 9, 22, 130, 133, 137, 140, 189
cross-pol, 26, 32, 37, 76, 95–96, 116, 130, 136–137, 140, 150, 168, 209, 222, 230

D

data mode, 92
detune, 5, 37, 46, 79, 205
dielectric constant, 4, 12, 24–26, 44, 46, 49, 55, 57–58, 74–77, 94, 116, 118, 131, 136, 143, 150, 164, 167, 176, 203, 205–206, 216–219, 222
diffraction, 37
discontinuity, 30–32, 45, 49, 57–58, 79–80, 143, 167, 204
dominant mode, 37, 83
dual-polarized, 150, 160, 214

E

effective radiating volume, 21, 46, 59, 73, 99–100, 157, 165, 181
end-fire, 11, 24, 41, 49, 65, 73, 76–98, 117, 119, 121, 127, 136, 139–140, 150, 153, 155, 177, 192, 197, 206, 222, 231–236
end-launch connector, 26, 28, 44, 46, 49, 57, 60, 76–77, 85, 93–95, 98, 103, 117, 141, 143, 181, 190, 195, 206–207, 229, 233–234
Envelope Correlation Coefficient (ECC), 170, 172, 237
equivalent circuit, 26, 76, 85, 89, 165–166

239

F

fabrication process, 24, 57–58, 73–74, 203–204, 219
feedline, 26, 28, 74, 77, 80, 85, 94, 96, 102, 143, 206
finger blockage, 11, 92–93, 102–103
flexible, 11, 53–71, 94, 164, 172, 176
folded dipole, 41, 74, 75–110
footprint, 7, 12, 14, 21–24, 37–38, 46–48, 49, 54, 57–60, 65, 74, 78, 80, 92, 94, 131, 138, 141, 149–150, 153, 160, 163–164, 165, 172, 189, 192, 194–195, 207, 214, 219, 222–225
frequency selective surface (FSS), 58
full-wave, 28, 59, 94, 229
fused deposition modelling (FDM), 216

G

gain, 4, 6–13, 21–40, 45–49, 54–70, 73–108, 115–145, 150–160, 163–185, 189–200, 206–236
gain enhancement, 12, 33–34, 40–41, 48–49, 65–66, 73, 82, 91, 100, 119, 138–139, 150, 153–154, 158–160, 189, 214, 220, 223
gain transfer method, 29, 78, 96, 137, 157, 181

I

impedance bandwidth, 12, 24, 26, 28, 31, 34, 37, 39, 41, 45–46, 58, 60, 67, 74, 76, 80, 82, 83–87, 91–94, 98, 100, 116, 118, 125, 133, 136, 139–140, 142–143, 151, 167, 177, 180, 194, 197, 214, 216, 222–225
indoor, 2, 8–9, 14, 74, 108–111, 117–120, 138, 145–146, 209
inkjet printing, 55
input reflection coefficient, 25–26, 28, 31, 48, 58, 60, 76, 85, 98, 100, 117, 125, 133, 167, 172, 177, 180–181, 197, 217, 237
insertion loss, 11–13, 115, 117, 139, 145, 153, 190, 193–194, 199–200, 206, 219
integrated, 4–5, 9, 11–12, 22–25, 38, 40–41, 45, 49, 59–60, 64–65, 73–74, 77, 81, 83, 85, 86, 89, 92, 94, 98–100, 103, 116, 119, 139, 140, 164–165, 176, 181, 184, 190, 192, 194, 214, 216, 219, 222

L

landscape, 14, 53, 54, 60, 74, 89, 92–93, 96, 100, 102, 108, 181, 189
leaky wave, 12, 121, 138, 151, 157
line of sight (LOS), 8, 210

link budget, 3–4, 6–7, 9, 22, 93, 98, 100, 108, 115, 119, 138, 156, 189
liquid crystal polymer (LCP), 55
log-periodic dipole, 12, 96
long term evolution (LTE), 2, 22, 163
loss tangent, 11, 25–26, 44, 49, 55–56, 76, 116, 131, 136, 150, 164, 176, 203, 206, 214, 216–217, 219, 229

M

manufacturing, 11, 23, 55, 60, 73, 92, 119, 167, 206, 215
matching circuit, 4, 78, 164, 200, 219
meandering, 164
metallic, 4–5, 11–12, 22, 50, 65, 108, 115, 178, 214, 223
metamaterial, 40, 49, 65, 89, 92, 100, 119, 121–122, 125–139, 145, 150, 153, 157, 160, 214, 223
mobile telephony, 4
mounting, 28, 87, 205, 207
multi-orientation, 11, 22
multi-port, 190, 207, 214
mutual coupling, 11, 22, 24, 26, 53, 62, 65, 89, 105, 116, 133, 142, 150–154, 189, 191, 196, 207, 214, 222, 237

N

noise floor, 108, 208–209
non-line of sight (NLOS), 8

O

offset, 37, 39, 78, 82, 85, 87, 100, 115, 124, 127, 136, 139, 142, 170, 190
on-chip antenna, 219–220, 222
orthogonal, 9, 11, 13–14, 21, 30, 32, 37, 45, 53–54, 58–59, 63–65, 79–81, 89, 92, 96, 100, 102, 105, 110, 118, 127, 149–160, 172, 181, 183, 185, 208, 214
overlapped, 54, 62, 65, 214
over-moded, 11, 26, 40, 77, 79

P

package, 12, 40, 204
panel, 4, 10, 21–22, 34, 54, 58, 94, 108, 163–164, 192, 213
parasitic elements, 11
parasitic ellipse, 99–105
path loss, 3, 5–9, 14, 29

path loss compensation, 116, 120–121, 126, 130, 132–133, 142, 145, 214
pattern integrity, 12–13, 45–46, 53, 60, 96, 98, 117, 127, 139, 151, 177, 181, 194, 222
penetration loss, 3–9, 149
perfect electric conductor (PEC), 28, 37, 77, 229
periodic boundary conditions, 37, 83, 121, 139
permeability, 119
permittivity, 119, 176
phase error, 108, 124, 139, 154
phase linearization, 124
phase shifter, 12, 64, 116, 149
phased array, 7–14, 22, 48–49, 51, 54, 115–116, 120, 138, 149, 151, 172, 189–201, 214
physical aperture, 7, 40, 49, 65, 89, 98, 124, 139, 151, 154, 236
polarization, 9, 22, 37, 49, 60, 65, 76, 83, 96, 108, 116, 121, 136, 149–160, 214
polycarbonate, 11, 55–59, 66, 73, 164, 176, 194, 214
polyethylene terephthalate (PET), 55, 164, 176
polylactic acid (PLA), 46, 131, 216
portrait, 53–60, 73–74, 89, 92–96, 108, 110, 145, 172, 181, 189
principal planes, 22, 28, 46, 95, 127, 139, 168, 230
printed circuit board (PCB), 11, 73, 190, 203, 213, 216, 219
printed ridge gap, 12, 92
progressive phase shift, 191, 199

Q

quarter-wave, 28, 33–34, 77, 82, 98, 117, 150, 196, 221
quasi-waveguide, 11, 37, 141

R

radiating aperture, 22, 26, 28, 30, 33, 34, 37, 39, 79, 82–83, 100, 102, 119, 122, 136, 139–140, 150, 154, 180, 189
radiation efficiency, 1, 4, 11–12, 49, 55, 92, 143, 163–164, 168, 176–177, 216, 219–222
radio frequency integrated circuit (RFIC), 219–220

real-world scenario, 14, 89, 115
reconfigurable, 11, 214–215

S

scaffolding, 21, 24, 30, 45–46, 49, 57, 62, 92, 92, 142, 164, 176
scanning loss, 11–12, 54, 64, 92, 138, 149, 151, 172, 190, 192, 200, 214
shared aperture, 149
side lobe, 116
slotline, 60, 98, 100, 117, 136, 143, 150, 154
specific absorption rate (SAR), 5, 33, 38, 74, 81
substrate integrated waveguide (SIW), 11, 24, 49, 64, 119, 190
superstrate, 12, 119, 131, 133, 149, 214, 235, 236

T

tapered slot antenna (TSA), 12, 49, 60, 96, 116, 136, 141, 143, 145, 150, 177, 205
tolerance, 24, 78, 75, 77, 126, 140, 150, 167, 169, 181, 203
transparent, 60, 131, 225

U

uniplanar, 30, 74, 79, 222
unit cell, 12, 34, 37, 40, 83, 121, 139, 153, 165, 214, 234

W

wavefront, 70, 124, 127, 139, 145, 154, 209

Y

Yagi, 11, 76, 117, 153

Z

zero-index metamaterial (ZIM), 136, 139–141, 150, 153–154, 156, 158